ISBN 978-1-332-59982-0
PIBN 10030593

English
Français
Deutsche
Italiano
Español
Português

www.forgottenbooks.com

Mythology Photography **Fiction**
Fishing Christianity **Art** Cooking
Essays Buddhism Freemasonry
Medicine **Biology** Music **Ancient
Egypt** Evolution Carpentry Physics
Dance Geology **Mathematics** Fitness
Shakespeare **Folklore** Yoga Marketing
Confidence Immortality Biographies
Poetry **Psychology** Witchcraft
Electronics Chemistry History **Law**
Accounting **Philosophy** Anthropology
Alchemy Drama Quantum Mechanics
Atheism Sexual Health **Ancient History**
Entrepreneurship Languages Sport
Paleontology Needlework Islam
Metaphysics Investment Archaeology
Parenting Statistics Criminology
Motivational

RESEARCH IN CHINA

IN THREE VOLUMES AND ATLAS

VOLUME TWO

SYSTEMATIC GEOLOGY

by

BAILEY WILLIS

WASHINGTON, D. C.

PUBLISHED BY THE CARNEGIE INSTITUTION OF WASHINGTON

JULY, 1907

CARNEGIE INSTITUTION OF WASHINGTON

PUBLICATION NO. 54 (VOLUME 2)

PRESS OF GIBSON BROTHERS,
WASHINGTON, D. C.

CONTENTS.

III

039

LIST OF ILLUSTRATIONS.

ERRATA.

Page 17, line 5, read 1903–04 for 1904–05
Page 49, line 24, read later for late.

INTRODUCTION.

The expedition to China carried out in 1903–04 under the auspices of the Carnegie Institution of Washington gathered certain observations relating to the geology of that country, which are stated in detail under the general subject of descriptive geology, in volume I, part I, of these contributions. The same material is treated in this volume, but from the standpoint of systematic continental history. Repetition has been avoided, so far as is consistent with the presentation of the general subject. Other sources of published information have been freely consulted and are referred to in connection with the discussions.

The subject-matter of this volume is classified primarily according to geologic eras or periods, and secondarily by broad areas of which China proper is the central region. Thus the reader will find that each chapter treats of the geographic condition of southeastern Asia during a particular age, and that successive chapters trace the sequence of changes from age to age. Only the great events of continental history are distinguishable on the basis of existing knowledge, but they suffice to show that the present continent is an aggregation of land masses which, from time to time, have been forced into union with one another. This generalization has already been stated by Suess, who, in cooperation with Neumayer, developed the recognition of the great mediterraneans which formerly divided Eurasia.

This study of Asia has been carried on with comparative studies of North America and Europe, and has led to theoretical views regarding continental structure and development which have a broad application. They are briefly stated in a closing chapter.

Many associates have helped me toward an understanding of the problems involved, and I desire to express my appreciation of aid received from geologists and physicists alike. An especial debt is due the two great leaders in the modern advance of geological philosophy, Chamberlin and Suess, from whose works and personal suggestion I have derived the highest inspiration. Their assistance is none the less earnestly appreciated because I have been compelled to a somewhat divergent view in theoretical questions of continental development.

BAILEY WILLIS.

WASHINGTON, D. C., *February 25, 1907.*

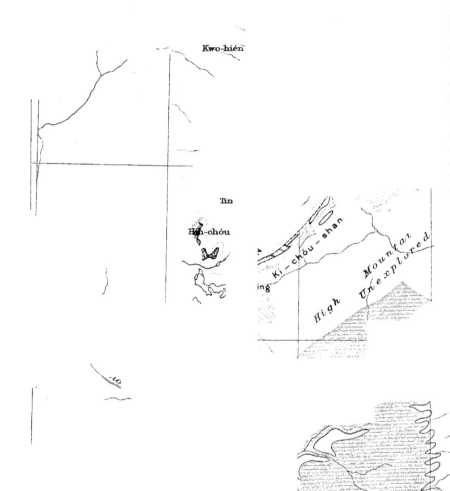

Kwo-hién

Tin

Hiah-chóu

Ki-chóu-shan

ing

High Mountain Unexplored

113°30′

Compiled from Surveys of
1903-04 and other Sources

Willis and Blackwelder
1903-04

F.von
Richthofen
1870

10   5

ROUTE

LEGEND

SEDIMENTARY ROCKS

PLATE 1

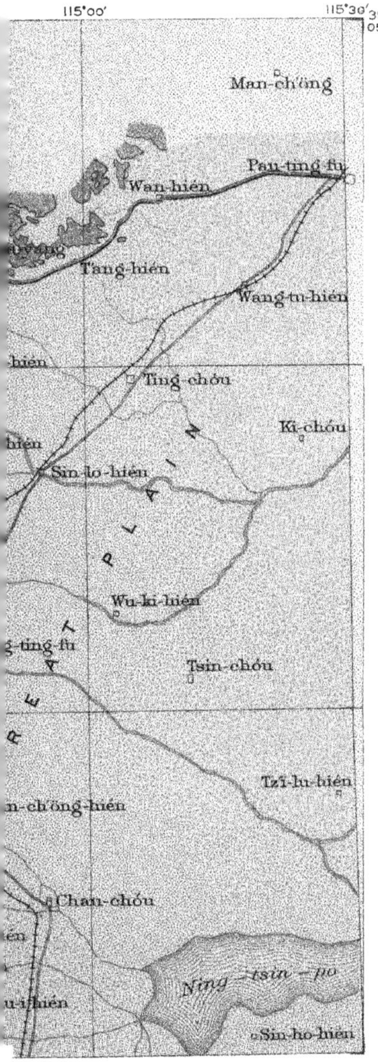

115°00'   115°30' 39°
05'

Man-ch'ōng

Wan-hién   Pau-ting fu

Tang-hién   Wang-tu-hién

hién   Ting-chóu

Ki-chóu

hién   Sin-lo-hién

Wu-ki-hién

-ting-fu   Tsin-chóu

n-ch'ōng-hién   Tzí-lu-hién

Chau-chóu

ién   Ning-tsin-po

u-i hién   Sin-ho-hién

Geology by Bailey Willis
and Eliot Blackwelder, 1903-04
and F. von Richthofen 1870

Huang-t'u formation
(chiefly loess with notable
amounts of sand and gravel;
also includes marsh and
pond deposits of limited
extent)

} QUATERNARY

UNCONFORMITY

Red Beds
(red shale and sandstone)

} PERMO-
MESOZOIC

Shan-si coal measures
(clays, shales, sandstones,
and coal beds)

} UPPER
CARBONIFEROUS

UNCONFORMITY

Sinian system
(gray to brown, thin bedded
and massive limestone
with thin beds of shale.

} CAMBRO-
ORDOVICIAN

UNCONFORMITY

Hu-to series
(chiefly gray limestone, interbedded with slate and
quartzite, or gray slate
with thin beds of siliceous
limestone, quartzite, and local conglomerate.

UNCONFORMITY

METAMORPHIC ROCKS

Wu-t'ai schists
(chlorite-schist with gray
schist and conglomerate
of quartz and quartzite;
siliceous marble quartz-
ite, and jasper, with interbedded biotite and chlo-
rite schists)

UNCONFORMITY

} PROTEROZOIC

Tai-shan complex
(gneiss and schist chiefly
of unknown origin, partly
sedimentary, with granite
and other intrusives)

} ARCHEAN

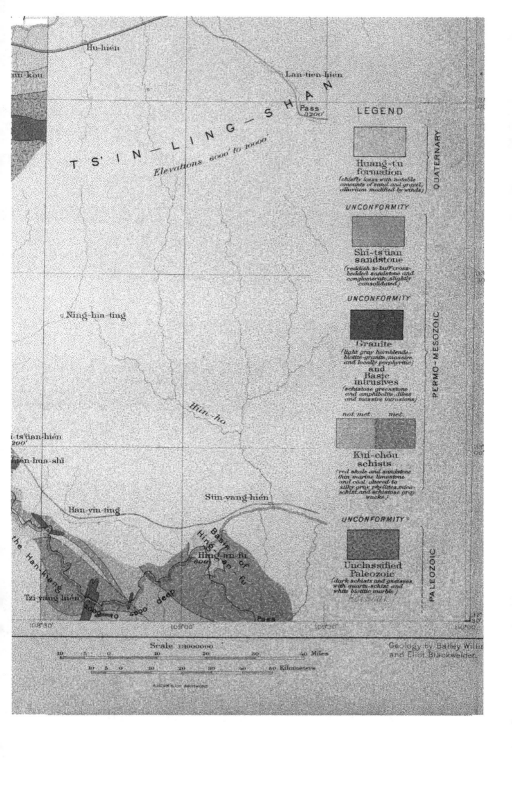

Hu-hien

-nu-kou   Lan-tien-hien

TS'IN—LING—SHAN
Elevations 8000' to 10000'

Pass
9200'

LEGEND

Huang-tu
formation
(chiefly loess with notable
amounts of sand and gravel;
alluvium modified by winds)

UNCONFORMITY

Shi-ts'üan
sandstone
(reddish to buff cross-
bedded sandstone and
conglomerate, slightly
consolidated)

UNCONFORMITY

Granite
(light gray hornblende-
biotite-granite, massive
and locally porphyritic)
and
Basic
intrusives
(schistose greenstone
and amphibolite, dikes
and massive intrusions)

not met.   met.

Kui-chóu
schists
(red shale and sandstone,
thin marine limestone
and coal altered to
silky gray phyllites, mica-
schist and schistose gray-
wacke)

UNCONFORMITY

Unclassified
Paleozoic
(dark schists and gneisses
with quartz-schist and
white biotite marble)

QUATERNARY

PERMO-MESOZOIC

PALEOZOIC

Ning-hia-ting

Hün-ho

i-ts'üan-hien
?00'

hien-hua-shi

Sün-yang-hien

Han-yin-ting

Basin of
Hing-an-fu

the Han Kiang

Hing-an-fu
800'

Tzi-yang-hien   2000'
1000' to   deep
Pass

108°30'   109°00'   109°30'   110°00'

Scale 1:2000000

10   5   0   10   20   30   40 Miles
10 5   0   10   20   30   40   50 Kilometers

Geology by Bailey Willis
and Eliot Blackwelder.

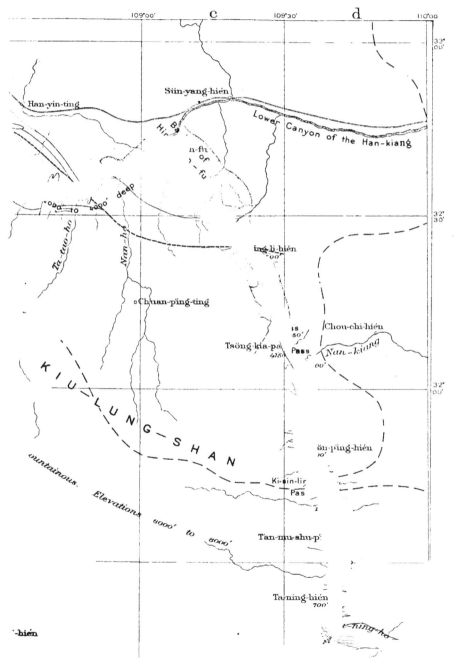

CHAPTER I.—ARCHEAN.

The term Archean has been used in Asiatic geology by different writers to cover various metamorphic but usually ancient rocks. Generally it is applied to any highly metamorphic or igneous rocks which are older than the oldest clearly recognized sediments of any particular district. In the literature "Archean" thus covers rocks which range in age from the very earliest pre-Paleozoics observed in the continent to Paleozoic, or possibly even Mesozoic strata, where these have been greatly altered. Moreover, the intrusives have commonly been grouped with the schists, although they are certainly somewhat younger, and it is sometimes difficult to say how much younger. Thus, granites which may be of Paleozoic or Mesozoic age are classed as pre-Paleozoic.

In these volumes the term Archean is applied to a basal mass of gneisses and schists, which is also designated the T'ai-shan complex. It is regarded as basal because it underlies all other recognized groups and is separated from them by an unconformity of a most distinct and profound character; moreover, it is fundamental among rock masses known at the surface, because it is itself apparently bottomless. Nothing distinct from it and older than it has as yet been identified.

The T'ai-shan complex is not everywhere followed by rocks of the same age, the geologic record being incomplete in China as elsewhere. In the Wu-t'ai district of northern Shan-si, it underlies the Shï-tsui group of the Wu-t'ai schists, probably the oldest group of sedimentary deposits identified in Asia. The most probable rough correlation places this sequence in parallelism with the lower Huronian and Keewatin of the Lake Superior region of North America.*

In northeastern China, in the province of Shan-tung, is the type locality of the T'ai-shan complex, Mount T'ai. The most ancient rocks are there overlain by Lower Cambrian strata and the pre-Cambrian sediments are wanting; correlation with the Archean of the Wu-t'ai district is based on similarity of lithologic and structural characters, which are common to all the occurrences which we have classed as T'ai-shan and are not shared by any other rock masses of known superior position in the geologic scale.

*Report of special committee for the Lake Superior Region, Chicago Journal of Geology, vol. xiii, p. 89, 1905.

The lithologic and structural characters of the T'ai-shan complex are those of the Archean of the Lake Superior region of North America, as defined by Van Hise. The rocks are chiefly metamorphic schists and gneisses of indeterminate original character; associated with them is a large proportion of metamorphosed igneous rocks and a very small proportion of metamorphosed sediments; and the metamorphics are intruded by granites, which are relatively young, though in large part probably pre-Cambrian.

The structure of the gneisses and schists is exceedingly intricate; it is characterized by a universal schistosity; by a common banding; by complex shearing, thinning, thickening, plication, and flow of bands; and such intricate arrangement of the very variable petrographic facies of the schists and gneisses as makes impossible any structural study by usual stratigraphic methods.

These lithologic and structural peculiarities of the T'ai-shan complex serve to distinguish it unmistakably from the oldest pre-Cambrian sediments, the Wu-t'ai system, even though the latter are intensely metamorphosed.

CHAPTER II.—PROTEROZOIC.

We recognize two principal systems of stratified rocks of the Protero-zoic era in northern China, the Wu-t'ai and the Hu-t'o or Nan-k'ou systems. The Wu-t'ai was observed and named by von Richthofen; the other he did not distinguish from the Sinian, of which he thought it a conformable part, and he described it as "Untersinisch." In certain sections and probably generally, there is an unconformity between the "Untersinisch" and the "Obersinisch," and it seems desirable to separate them by restricting the term Sinian to upper or Obersinisch and giving a new name to the lower or Untersinisch.* Nan-k'ou is an appropriate name, which may be applied if we correctly understand von Richthofen's section at the celebrated pass northwest of Peking. Hu-t'o is a name we gave to a series which is probably the equivalent of the Nan-k'ou, though it differs somewhat in the character of the strata. The matter is discussed in a later paragraph.

Proterozoic rocks are best exposed in China, so far as yet known, in a mountain range in northern Shan-si, called the Wu-t'ai-shan. This name, which means Five Platform Mountains, is derived from the five highest summits in the region, Peï-t'ai or North Table having an altitude of 10,045 feet. The range is deeply incised, and the canyons expose remarkably distinct sections of ancient metamorphic rocks, comprising the basal or T'ai-shan complex (Archean) and the Wu-t'ai system, eo-Proterozoic. The latter is thought to be roughly correlative in age and duration and complexity with the Huronian of Canada.

On the southeastern flank of the Wu-t'ai-shan is a wide area of hills and valleys, which we have called the Northern Loess Basins. The general altitude is 3,500 to 5,000 feet above sea. Rocks of the early Proterozoic, Wu-t'ai, system constitute part of the surface, but a younger pre-Cambrian system of but slightly altered quartzites, slates, and siliceous limestones covers a larger area. This neo-Proterozoic system we have called the Hu-t'o, a local name, which may ultimately be replaced by the term Nan-k'ou of von Richthofen or that of Nan-shan of Lóczy.

In view of the better, though still incomplete, knowledge which we have of the Proterozoic in the Wu-t'ai as compared with notes of occur-rence of similar rocks elsewhere, we may take the Wu-t'ai sections as types.

---

* In a personal conference in Berlin, March, 1905, von Richthofen agreed orally to the desirability of making this distinction, but his views were not put in writing, so far as I am aware.

WU-T'AI SYSTEM (EO-PROTEROZOIC).

Rocks of the Wu-t'ai system constitute much of the Wu-t'ai-shan, a short but high mountain range in northern Shan-si, longitude 113° 30′ E., latitude 39° N. Von Richthofen crossed the mountains in December, 1871, and clearly observed the existence of a series of strata which he had not previously seen in the course of his extended journeyings; nor are they elsewhere known in equal mass. During the stormy December weather which he met the conditions were most unfavorable for observation, and his route crossed only a part of the Wu-t'ai rocks, that mass of chlorite schists which we consider the upper part and call the Si-t'ai formation. The lower series of mica schists, quartzites, and marbles he did not see.

Von Richthofen placed the Wu-t'ai system in the Huronian, using the latter term, as was commonly done thirty years ago, to suggest pre-Paleozoic metamorphic rocks of green color.* He does so with reserve, however, and the stricter usage of the term as it is now adopted† does not permit us to maintain an exact correlation. Yet there is a certain parallelism in stratigraphic position and history, as well as likeness in rocks, which invites comparison of the Wu-t'ai and the Huronian, as will appear in stating the facts as we now understand them.

Our own opportunities for study of the Wu-t'ai system were limited practically to a single section and a few isolated localities. In volume I we have stated our inference regarding the stratigraphic sequence and structure as provisional only, and we commend to some future investigator the almost unequaled exposures of ancient sedimentary and igneous rocks which represent one of the earliest intelligible records of Asiatic history.

A provisional classification of the Wu-t'ai and limiting systems in the type locality is as follows:

Hu-t'o system (Neo-Proterozoic)	Tung-yü limestone } Tóu-ts'un slates	Slates, limestone, and quartzite.
	Unconformity	
Wu-t'ai system (Eo-Proterozoic)	Si-t'ai series	Chiefly chlorite schist; quartzite conglomerate at the base.
	Unconformity	
	Nan-t'ai series	Siliceous marble, jasper, quartzite, and schist.
	Unconformity	
	Shï-tsui series	Mica schists, gneiss, magnetite quartzite, and basal feldspathic quartzite.
	Unconformity	
T'ai-shan complex (Archean)		Basal complex of varied gneisses and younger intrusives.

A brief account of the successive elements of the Wu-t'ai section follows (see atlas sheet DI and plate XVIII, volume I, part I).

*China, vol. II, p. 377.
† Report of International Committee, Chicago Journal of Geology, vol. XIII, 1905.

That there is an unconformity at the base of the Shï-tsui series, separating it from the T'ai-shan complex, we do not doubt. The distinction
between the stratified schists and quartzites of sedimentary origin and the
very intricate complex of parallel bedded injection gneisses is decided when
they are contrasted in extensive sections. But the exact contact between
the two is not readily recognized. In a ravine south of the T'ai-shan-ho
we found a peculiar quartzite containing large crystals of red feldspar,
which is a recrystallized arkose, occurring at or near the base of the Shï-tsui
series, on or close to the T'ai-shan gneiss, from which the material was
probably derived by erosion. The immediately underlying stratum is a
mica schist of indeterminate origin. The feldspathic quartzite grades
upward through quartzite containing some mica and but little feldspar,
into mica schists, which present great variety of mineralogical composition
and are in turn succeeded by quartzites which are in part magnetic. The
section of sediments, which has a length of eight miles, is interrupted by
a mass of augen-gneiss that is probably intrusive. The dip of the Shï-tsui
strata in this section is from 30° to 70° to the northwest, and the repetition
of the quartzite on two sides of a great body of mica schists is such that
the probable structure is a syncline overturned toward the southwest.

The augen-gneiss above referred to extends for three miles along the
T'ai-shan-ho. It has the uniformity_of composition of a batholite and
appears to be an intrusive which has suffered shearing and metamorphism
with the sedimentary Wu-t'ai rocks. Northwest of it occur the strata of
the Si-t'ai and Nan-t'ai groups.

Nan-t'ai and Si-t'ai are two of the five peaks of the Wu-t'ai-shan,
each of which is composed chiefly of the strata named after it.

Strata of the Nan-t'ai group are siliceous marble, gray to black or
red quartzites, and schists, chiefly of chlorite and muscovite. They are
also garnetiferous and staurolitic. They occur on the southeast slope of
the Wu-t'ai range and are well exposed in the canyons above Shang-ho-
miau. The structure shows two or more closed synclines, overturned
toward the south, and overthrusts which eliminate the anticlines. The
sequence, as we interpret it, consists of siliceous and muddy sediments below
and calcareous deposits above. It may be continuous with or distinct
from the Shï-tsui. The two groups of rocks were not seen in contact.

The Si-t'ai group, comprising most if not all 'of the sequence seen by
von Richthofen, forms the mass and northern slope of the Wu-t'ai-shan.
It is a great body of green schists, chiefly chloritic, with beds of quartzite
containing magnetite, and coarse conglomerate of quartz and quartzite
cobbles at the base; all as he described it.* The relations with the Nan-
t'ai group are well exposed in the canyons on the south slope of the range.

---
*China, vol. ɪɪ, p. 364.

We there find the conglomerate duplicated by overthrust, but overlying in regular sequence the Nan-t'ai quartzites, and grading from conglomerate through quartzite upward to schist, after the fashion of a passage from basal conglomerate to fine muddy sediments. The pebbles in the conglomerate consist chiefly of quartz and quartzite, the latter being of varieties that occur in the Nan-t'ai group, together with some of granite. Hence we infer that the rocks below the Si-t'ai had been folded and eroded before that series was deposited. The conglomerate, consisting as it does almost entirely of large rounded quartz and quartzite pebbles, contains the remains of a deposit that had been concentrated from a more complex constitution. Before metamorphism it may have closely resembled the bed of quartz cobbles found at the base of the Potomac formation of the Atlantic coast, and may have resulted, as that deposit did, from a prolonged cycle of erosion which ended in consequence of a marine transgression across a peneplain. Moreover, in the western part of the Wu-t'ai-shan, no groups corresponding to the Shï-tsui and Nan-t'ai groups were seen between the T'ai-shan complex and Si-t'ai schists. This fact and the presence of the granite pebbles in the conglomerate suggest an overlap of Si-t'ai strata beyond the older Proterozoic rocks.

The greater part of the Si-t'ai group consists of green schists colored chiefly by chlorite. Stratification is obscured by schistosity and the structure can not readily be made out. It is probable that several isoclinal folds occur, overturned toward the south and separated by overthrust faults.

On the summit of Peï-t'ai is a small area of biotite gneiss, which may be the highest stratum of the Si-t'ai group preserved in a syncline, but we were not able to determine conclusively what the relations to the adjacent green schists actually were. The gneiss may be an intrusive body.

Returning to the analogy which exists between the Wu-t'ai and Huronian systems, I may point out some of the parallel relations. The Lower Wu-t'ai (Shï-tsui group) rests unconformably upon the basal T'ai-shan complex, as the lower Huronian does on the Archean (Keewatin) gneiss. The Nan-t'ai group overlies the Shï-tsui and may be separated from it by an unconformity, as the middle Huronian is from the lower. The two series were similarly composed of siliceous, clayey, and ferruginous sediments, which have undergone intense metamorphism and become schists of varied constitution. The upper Wu-t'ai (Si-t'ai group) is unconformable to the middle Wu-t'ai as the upper Huronian is to the middle Huronian. The upper group in each case consists largely of chloritic schists associated with ferruginous quartzites. [Both the Wu-t'ai and the Huronian series have been affected by some igneous intrusions, which occurred before the strata were deformed by shearing, and also penetrated by later dikes. These are but analogies, yet they serve to suggest a parallel in the ancient history of Asia and North America, which extends to successive events of erosion, deposition, and deformation. The general relations to the Archean and neo-Proterozoic are similar in both continents, and the effects may well have been due to a general terrestrial cause which became active at about the same times, in regions remote from one another.

Before discussing the occurrence elsewhere in Asia of pre-Cambrian rocks, which may be equivalent to the Wu-t'ai, I describe the next younger, the Hu-t'o system.

HU-T'O SYSTEM (NEO-PROTEROZOIC).

All of the rocks of the Hu-t'o system are sedimentary strata: conglomerate, quartzite, shale, and limestone, which resemble the unmetamorphosed Paleozoic rocks more nearly than they do the Wu-t'ai schists. The physical events which intervened between the close of the Wu-t'ai period and the beginning of the Hu-t'o involved greater changes and probably longer time than those which occurred 'after the Hu-t'o and before the Sinian; but the presence of a rich fauna in the Sinian seas distinguishes that period from the preceding time, during which the life forms, though probably numerous, did not generally become fossil. The nearest relations of the Hu-t'o system are with the Belt terrane of Montana, and it is probable that pre-Cambrian fossils* such as have been found in the Belt may eventually be discovered in the Hu-t'o.

The Hu-t'o rocks occur in typical development in the district of Wu-t'ai-ihén, in northern Shan-si, where they form the southwestern part of the Wu-t'ai-shan and the hills about the Northern Loess Basins as well as along the Hu-t'o river. They occupy a broad synclinorium and lie between the Wu-t'ai schists and Sinian strata, in unconformable relations to both systems. Although the contact with the underlying Wu-t'ai series must be extensively exposed in the mountains southwest of the upper T'ai-shan-ho and also north of the Loess Basins, we did not see it. It was covered by loess in each of the sections along which we crossed it, and we did not recognize the distinct position of the unfamiliar Hu-t'o rocks in time to search for the contact. There can, however, be no doubt of an unconformity between the schists of the Wu-t'ai and the little-altered, slightly slaty beds of the Hu-t'o. After the Wu-t'ai sediments had been deposited they were folded, depressed by folding or subsidence to a notable depth, intruded by large igneous masses, and deformed under great pressure, so that their original structure was replaced by schistosity and their constituents were recrystallized. The Hu-t'o rocks suffered none of these changes. They were deposited only after the schists had been exposed by uplift and deep erosion. Obviously an interval of the first magnitude intervened.

---

*Pre-Cambrian fossiliferous formations, C. D. Walcott, Bull. G. S. A., vol. x, p. 199, 1899.

Observations of the contact between the Hu-t'o and the overlying Sinian show clearly that the former were folded, intruded by dikes, and eroded, before the latter were laid down.*

The Hu-t'o strata apparently constitute a sequence, which consists of conglomerate and sandstone, clay slates, and siliceous calcareous strata. The entire series is thinly bedded and the three elements (sands, clays, and carbonates) occur interbedded. In the type locality we observed a number of partial sections† which we could not connect on account of the extensive covering of the Huang-t'u, the loess formation. The sandy rocks appear to preponderate near the base and the limestones increase in proportion toward the top. This distinction appeared to be sufficiently marked to separate a lower argillaceous group from an upper calcareous one, and we have described the former as the Tóu-ts'un slates and the latter as the Tung-yü limestone. The division has local significance only.

The Tóu-ts'un slates comprise the lowest beds of the Hu-t'o series and have a thickness of not less than 3,500 feet, 1,000 meters; they may reach 5,000 feet, 1,800 meters. The principal strata are earthy gray to purplish slates, or locally phyllites, with which occur thin layers of buff to pink dolomite and siliceous limestone. The lowest beds seen, which we take to be near the base of the series, are of red and gray quartzite with local layers of conglomerate.

The Tung-yü limestone includes all the upper part of the Hu-t'o system. The characteristic rock is gray limestone, in which chert is usually present in notable quantity and often in masses or sheets. The massive beds of limestone vary in thickness from 10 feet to 500 feet, 3 to 150 meters, or more, and strata of gray to red slates or phyllites occur between them. The sequence is very like that of the underlying Tóu-ts'un group, but the relative proportions of slate and limestone are reversed. The thickness of the Tung-yü limestone is probably 3,000 feet, 900 meters, or more.

Throughout northern China there is a limestone formation which is lithologically identical with the heavy beds of the Tung-yü group, and occupies the stratigraphic position of the whole Hu-t'o series. We regard it as the equivalent of the Hu-t'o, but have given it the distinctive name of Ta-yang from a village in Chï-li near which it is well developed, longitude 115° 50′, latitude 38° 45′.

The Ta-yang is a dark-gray, massive limestone, which is distinguished from the very similar Sinian limestones by abundance of chert and absence of fossils. The formation is exceedingly uniform in lithologic character; the greater or less proportion of chert and the occasional occurrence of white quartzite being the chief variations. A notable characteristic is found

---

* Vol. I, p 136.
† *Ibid.*, pp. 123–125.

in curved and minutely contorted laminæ of flint, the contortion of which appears to be independent of local deformation of the strata. Similar bands have been noticed in the Siyeh limestone of the Belt formation, Montana.* They are noted by Lóczy as occurring in the Nan-shan sandstone of Tibet.† Similar forms, described by Stose in Cambrian limestones of the Appalachian Valley, are classed as *Cryptozoan proliferum* Hall.‡ Whether the peculiar structures are in some cases mechanical or organic is not known, but they have a long range—from late Proterozoic to Ordovician. They have not, however, been seen in the Sinian. Neither have the oolitic and conglomeratic phases of the latter been observed in the Ta-yang.

The thickness of the Ta-yang in a partial section measured by Blackwelder west of the type locality exceeded 1,200 feet, 360 meters, by an unknown amount. Willis observed a section northwest of T'ang-hién, Chï-li, which, though interrupted by at least two normal faults, appeared to include 6,000 feet, 1,800 meters, of the limestone.§ The thickness is, no doubt, several thousand feet.

At the base of the Ta-yang limestone is usually a thin stratum of slate or phyllite, which rests in marked unconformity upon the T'ai-shan complex (Archean). No doubt the limestone may occur in unconformity on the Wu-t'ai schists, as does its probable equivalent, the Hu-t'o system, but we did not observe any such instance. The occurrences we saw were at points west of Pau-ting-fu, Chï-li, and have been described in detail.‖

A contact of the Ta-yang limestone with strata of Sinian age was observed at Nan-t'ang-meï, Chï-li (see vol. 1, fig. 27, and atlas sheet F I). A very heavy bed of residual chert pebbles, derived from the pre-Cambrian limestone there, occurs beneath quartzite and black argillite, which are conformable to the overlying Sinian and which, though lithologically peculiar, are believed to belong to that system. The body of residual chert is regarded as an accumulation on the eroded surface of the Ta-yang, which was therefore folded and exposed before the lowest Sinian stratum was deposited.

Having thus described the Ta-yang limestone and its relations as we saw them, I give the following sections from von Richthofen, in which he noted strata that closely resemble it in character and stratigraphic position and which we consider to be its equivalents. These sections also include the Sinian, namely, von Richthofen's Obersinisch, and the Ta-yang equivalents are described by him as the Untersinisch. He did not

---

*Stratigraphy and Structure of the Lewis and Livingston Ranges, B. Willis, Bull. G. S. A., vol. XIII, p. 305, 1902.
† Reise des Grafen Széchenyi in Ostasien, vol. I, p. 553.
‡Chicago Journal of Geology, vol. XIV, pp. 210 and 217, 1906.
§Fig. 27, vol. I, page 131.
‖Vol. I, p. 130.

observe an unconformity between the two, but our observations of other sections, nevertheless, lead us to infer that an unconformable relation is general between the Ta-yang and the Sinian systems.

Von Richthofen's first section is across the Nan-k'ou range, northwest of Peking, en route to Kalgan.

SECTION IN THE NAN-K'OU RANGE (VON RICHTHOFEN).*

Description of the rocks.	Thickness (feet).	Equivalents in terms of this volume.
m Evenly bedded limestones without chert, dark gray to blackish; fine-grained; break splintery and conchoidal, sometimes uneven. Beds 2 to 12 inches thick, seldom more, well separated from one another. Non-fossiliferous. Here occur the limestones with horizontally embedded, flattish, rounded bodies which give worm-like outlines in cross-section. The bodies here consist mostly of dense black, the matrix of crystalline limestone. These "Wurmkalke" are everywhere characteristic of an upper horizon. Another typical rock is greenish splintery limestone. Total thickness.......	2000	The limestone (m) comprises strata which are characteristic of the Tsi-nan and Chau-mi-tién limestones. The non-fossiliferous condition is specially marked in the former, but fossils are not common in the latter. The "Wurmkalke," conglomeratic limestones, are well developed at the base of the Chau-mi-tién. A distinction between the two would scarcely be recognized in reconnaissance. The thickness is less than their combined volume, but the upper part of von Richthofen's section is covered.
l Globulitic limestones, clear gray to black. Globulites mostly size of oat kernels, seldom as large as peas. Beds 2 inches to 2 feet thick. Trilobites abundant.....	500	Strata comprised under i, k, and l correspond to the oolitic limestones and green shales of the Kiu-lung group of Shan-tung. The thickness of oolitic limestone is four
k Green strata......	80	times any measure which was there seen. It may be overestimated, but if correct
i Globulitic limestones like l, predominating, with dense homogeneous limestones interbedded..................	1500	does not affect the probability of close correlation.
h 5. Red clay shales..................120' 4. Gray limestone .... 80' 3. Red strata... .... 80' 2. Dense siliceous limestones of flat conchoidal fracture, very evenly bedded; whitish, greenish, reddish; prevailingly greenish-white .200' 1. Red clay shales..................100'	580	Beds 1 to 5 inclusive (h) are very characteristic of the Man-t'o formation of Shantung. The 200-foot limestone bed is much thicker than any occurring there, but holds a stratigraphic position like that of a thin layer in the type section.
g Gray, dense limestone, in part fine-grained crystalline..................	400	Beds f and g are unfamiliar; they appear to be upper members of the Ta-yang limestone, which were eroded in the sections we observed.
f Black and gray sandy clay shales and sandy, in part micaceous slates, partly even, partly curved. (Krummschalig) Yellow sandstone with iron ore........	500	
e Whitish-gray, fine-grained, crystalline limestone, with interrupted interbedding and knots of black flint...	2500	Beds d and e are typical Ta-yang limestone, and in view of the great thickness, the lithologic identity, and similar position below the Man-t'o, we feel confident of their being equivalent to that formation.
d Blue crystalline siliceous limestone, in part pure, in part interbedded with thick beds of flint. Thickness not determined, but great.		

*China, vol. II, p. 306.

The close comparison practicable between the observations which von Richthofen made in rapidly crossing the Nan-k'ou pass and the detailed sections we were able to measure in Shan-tung bears testimony to the accuracy and painstaking character of his work. The differences are not greater than might be expected in sections 300 miles apart, and all the characteristic earmarks of the formations are noted by him in their appropriate succession. Although he was not able to collect fossils, he observed their occurrence at horizons elsewhere abundantly fossiliferous. That von Richthofen did not see any unconformity between the Sinian and the strata *d, e, f, g,* which we assign to a pre-Sinian system, is not surprising. He remarks that the dip in the red strata (basal Sinian) was variable. The limestones are very much alike in the two systems, and even one who is looking for an unconformity may easily pass the contact in crossing a single section en route to "distant night quarters which were difficult to reach." Negative evidence of that kind has little or no weight. Nevertheless, it is obvious that the strata in the Nan-k'ou pass are less closely folded than in the vicinity of Si-ta-yang, where we observed the pre-Sinian rocks, and it is possible that an unconformity of dip may not exist. But even if the strata be conformable, it would still, in our judgment, be undesirable to extend the term Sinian to the great body of siliceous limestones below the Man-t'o shale, the first formation laid down in consequence of the transgression with which the Sinian period opened.

If the equivalency of the Ta-yang limestone with that exposed in the Nan-k'ou pass be eventually established, it will be desirable to abandon the term Ta-yang for Nan-k'ou as the name of the system. Nan-k'ou is the locality where the strata were first distinguished; they are apparently more completely represented there. The locality is at once well known and accessible, it being on the highway from Peking to Kalgan. Believing the strata to be equivalent, I regard Ta-yang as a local name that in systematic discussion should give way to Nan-k'ou. Cherty limestones of the Nan-k'ou terrane are described by von Richthofen as outcropping in a ridge adjoining the Nan-k'ou range,* but the section is incomplete.

In southern Chï-li, near "Hwo-lu-hsien," longitude 114° 30', latitude 38° +, a section of limestones is exposed in the "Hsi-p'ing-shan" mountains bordering the Great Plain, which von Richthofen describes as follows from above downward:

Firm sandstone, white and gray.
Crystalline limestone.
(Interruption.)
Greenish gray, brightly colored, ringing. thin-bedded siliceous limestones; interbedded with various sorts of strata.

---

*China, vol. II, p. 343.

Globulitic and conglomeratic ("Wurmkalk") limestones.
Thin-bedded limestones, red and green.
Red shales.
 (Interruption.)
Crystalline limestone with nodules and layers of flint.
 (Interruption.)
Alternation of crystalline, thin-bedded, limestone, including thin layers of flint, with
 quartzite, epidote rock, red sandstone, etc.
 (Interruption.)
Gray crystalline limestone.

This section does not present sufficient continuity to enable us to identify the several formations precisely, but the sequence of three typical groups of strata, cherty limestone, red shale, and characteristic Sinian limestone, is apparent. The cherty limestones are evidently the same as the Nan-k'ou terrane.

With reference to the locality at which we observed the Ta-yang limestone, the two sections cited from von Richthofen lie respectively, the Nan-k'ou northeast, distant about 200 kilometers, and the Hsi-p'ing-shan southwest, 80 kilometers. The Ta-yang locality thus lies between the two, and all three occur in the foothills that bound the Great Plain on the west. The greater part of the Nan-k'ou system in this northeast-southwest trend is siliceous limestone, which, from its extent and unusual thickness, is obviously a marine deposit.

Both von Richthofen and Blackwelder have observed the older sedimentary rocks of Liau-tung. After describing the fundamental gneiss the former sums up his observations* by enumerating two groups, namely, black quartzites and hornblende schists, which are intruded by granites and greenstones and weather down to a rich soil and rolling landscape; and the Ta-ku-shan series, consisting of firm, well-stratified quartzites of yellowish and whitish tints, together with clay schists, mica schists, and crystalline limestone. The schists and limestone occur in intimate but indeterminate relations with the quartzites. The schists attain a thickness of several thousand feet; the limestone is of limited or local occurrence only. These strata are intruded by the Korea granite, and unconformably overlain by a red cross-bedded sandstone with occasional conglomeratic layers, the Yung-ning sandstone.

According to von Richthofen, the Yung-ning sandstone belongs to his "Untersinisch."† Blackwelder,‡ on the other hand, regards it as a local development of the Man-t'o shales, the red littoral deposit that marks the

*China, vol. II, p. 106.
† Ibid., vol. II, pp. 73 and 109.
‡ Research in China vol. I, p. 95.

Sinian (Cambrian) transgression; and if that be the case it falls into von Richthofen's "Obersinisch."

In Shan-tung von Richthofen distinguished a metamorphic series which included limestone and certain quartzites. The two were not seen in contact and their relations remain indeterminate between themselves as well as to similar metamorphic formations elsewhere. The sequence, which includes crystalline limestone, occurs in eastern Shan-tung, between Chefoo and Töng-chóu-fu, with a thickness of several thousand feet. The lower part is mica schist, which, higher in the succession, gives place more and more to alternating limestone beds which attain great thickness and make up the upper part of the system in the "Kung-sun-shan." This stratigraphic sequence resembles that of the lower Wu-t'ai series of Shan-si, and the antiquity of the rocks, as judged by schistosity and metamorphism, is similar; but the series may be of some other Proterozoic age.

The quartzites occur near Ch'ang-kiu-hién, in eastern Shan-tung, and constitute a mountain mass cut through by diorite. They are much folded in a region where the Sinian is not, and are of pre-Sinian age, probably equivalent to the Ta-ku-shan quartzites, according to von Richthofen.

PRE-SINIAN ROCKS OF CENTRAL CHINA.

The typical occurrences of rocks of the Wu-t'ai and Hu-t'o or Nan-k'ou systems are confined to the eastern part of the continent. They lie 2° to 4° of longitude west of the present eastern coast, between the parallels of 38° and 40° north. In the same latitude they are 24° to 26° east of the central meridian of Asia, which we may take as longitude 90° east. They are apparently isolated areas, known only in the mountainous region of northwestern China, and may eventually prove to be provincial systems, which can not be precisely correlated with terranes of other regions. A similar condition exists in North America, where rocks of Proterozoic age are known in several widely separated districts, but they have not been more closely correlated than as earlier and later Proterozoic. Nevertheless, rocks having the lithologic characters of the Wu-t'ai and Nan-k'ou strata and holding a somewhat similar position between those which are classed as Archean and deposits which are identified as Paleozoic, are known elsewhere in Asia, and it is desirable to place them roughly in parallelism.

Our own observations are limited to a single section of the Ts'in-ling-shan, Shen-si, and the outcrop in the lower Yang-tzï gorges, Hu-peï.* Von Richthofen,† Lóczy,‡ and Obrutchov§ are the original observers upon

---

*Vol I, pp. 265 and 313.
†China, vol. II, p. 557 et seq.
‡Reise des Grafen Széchenyi in Ostasien, vol. I, chapters VII, VIII, and IX.
§Northern China and the Nan-shan (in Russian).

whom we must chiefly depend. The citations from Russian geologists, given by Suess,* are very valuable.

I take up the occurrences of rocks which may be related to the Wu-t'ai and Nan-k'ou systems, in geographical sequence, proceeding from the nearer to the more remote. The typical terranes of the pre-Sinian sediments in North China are found in the Wu-t'ai-shan, longitude 114°±, latitude 39°±, northern Shan-si, and in the Nan-k'ou range, longitude 116° ±, latitude 41° ±, northwestern Chï-li. They have been described in the preceding section.

In central and southern Shan-si, quartzites and schists of typical pre-Sinian character occur in the mountains northwest of Fön-chóu-fu, longitude 112°, latitude 37° +, in the Ho-shan, longitude 112°, latitude 36° to 37°, and in the Föng-huang-shan, longitude 110° to 111°, latitude 35°. They appear on anticlines or normal fault-scarps, below Sinian strata. Our information regarding them is limited, however, to identification of float brought down by streams and distant observations of their characteristic outcrops beneath limestone scarps, as neither von Richthofen nor any of his successors has seen them at close range in place. Our notes are given in volume I, page 171 *et seq.*

A great mountain chain, the Ts'in-ling-shan, stretches from east to west across Central China, between the meridians of 104° and 114° east, about 700 kilometers southwest of the Wu-t'ai-shan. This barrier range, which is regarded as the eastern continuation of the Kwen-lung, consists chiefly of Paleozoic and Mesozoic strata. It is the northern margin of a geologic province that is distinguished by metamorphism of the Fermo-Mesozoic and older sediments, which in northern China are unaltered. In this respect it bears to that region much the same relation as that which the Sierra Nevada of California has to the Great Basin province of Nevada and Utah. Within this metamorphic province there occur quartzites, phyllites, and even schists, which are in fact of Sinian age or younger, but which have pre-Sinian aspects. We are therefore obliged to be cautious in making correlations on a lithologic basis.

The eastern end of the Ts'in-ling-shan is described by von Richthofen under the name Fu-niu-shan, and a northeastern outlier of imposing altitude is the Sung-shan. These heights are in Ho-nan, south of the Huang-ho, longitude 113° east, latitude 34°±. In describing the sections which he saw in the Fu-niu-shan,† von Richthofen mentions chloritic schists and crystalline limestone; less distinctly crystalline and gray-green slates (Thonschiefer) and slaty quartzites; and these strata, which are steeply folded, are unconformably overlain by coarse conglomerate and sandstone

---

* Face de la Terre, vol. III.
† China, vol. II, pp. 496–497.

in which coal-beds occur. This is the section in the southern ridge, the Kiu-li-shan. A range which succeeds on the north consists of crystalline limestone, crystalline schists, gneiss, and a large mass of granite. The granite rises 2,500 to 3,000 feet, 750 to 900 meters, above the narrow pass; it consists of a ground-mass of orthoclase, greenish plagioclase, much quartz, some brown mica, and sporadic hornblende of medium texture, in which occur crystals of bright flesh-red orthoclase, that have the form of Carlsbad twins and attain a diameter of 4 inches.

The granite is plainly intrusive in the gneiss, which it traverses in large dikes. Quartz veins are also extensively developed. Crystalline schists intruded by quartz porphyry complete the section of the range on the north. Oolitic limestone of Sinian character, but which seemed to be unfossiliferous, occurs in a parallel ridge a short distance to the north and is unconformably overlain by coal-bearing sandstones. The latter are intruded by "greenstones" and porphyritic eruptives. Von Richthofen maps the crystalline schists and associated rocks of the region as "gneiss and crystalline schists in general," with the color with which he indicates the pre-Sinian (Archean) basement. The assignment to a pre-Sinian age is borne out by the apparent relation between the metamorphics and the unaltered strata classed as Sinian. In the grouping which is adopted in this report the predominantly sedimentary rocks of pre-Sinian age fall into the Proterozoic and their lithologic associations are with the Wu-t'ai system, or possibly with the Hu-t'o. A qualification of this inference lies in the intrusive character of the granite and its possible Mesozoic age. If the metamorphism is the effect of an episode of deformation with which the intrusions are related, and the granite is post-Carboniferous, then the schists, quartzites, and crystalline limestones may be Paleozoic. There is, however, nothing to sustain this qualification except the fact that post-Paleozoic granites occur in the Ts'in-ling-shan. Although not in the least gneissoid, and therefore apparently young, the intrusive is in this respect, as well as in petrographic constitution, identical with the Korea granite of Shan-tung and Liau-tung, which von Richthofen determined to be pre-Sinian* and with which he compares it.

The Sung-shan in Ho-nan, one of China's five holy mountains, was seen by von Richthofen from a distance. It is an isolated mass estimated by him at 8,000 feet altitude. A principal peak, the Yü-tsai-shan, exhibits the form common to masses of the coarse granite, while the main range shows those sculptured from crystalline schists. A relatively low ridge, the "Hsiung-shan," on the south of the Sung-shan, appeared to the traveler to consist of lower and upper Sinian strata resting on a base of schists.

*China, vol. II, p. 83.

Between the meridians 112° and 109° east, Lóczy crossed the Ts'in-ling-shan on the highway which connects the valley of the Han with that of the Weï. In ascending the "Sié-ho" (Siau-ho or Little River) he noted several sections of Paleozoic strata, which were not metamorphosed, rising between basins of Mesozoic deposits. On the north he next observed graphitic schists, pyritiferous quartzite, and yellow dolomite, which he classes as "submetamorphic Paleozoic schists." The description of them* corresponds closely with rocks which we observed on the Han and south of it, between longitude 108° 30' and 109° 30' east, and classed as Middle Paleozoic and Carboniferous. The observations thus agree.

After crossing the metamorphosed Paleozoic strata, Lóczy came upon a broad belt of biotitic schists of monotonous character, with which occur dark amphibole schists and gneiss, chlorite gneiss, and lenses of white granular limestone. These metamorphosed sediments are intruded by massive diorite, coarse-grained amphibole granite, and pegmatite, which locally change the schists to hard, fine-grained gneisses. He classes the complex under the Azoic or Archean formations. They are unconformably overlain by slightly coherent conglomerate, sandstone, and shale, of Jurassic age; beyond these superjacent beds reappear quartzitic dolomitic limestone, quartzite, and gray micaceous phyllites of the metamorphic Paleozoic group; and they are in a short distance again succeeded by a broad zone of monotonous mica schist, amphibole schist, and gneiss, penetrated by eruptives. The main range of the Ts'in-ling-shan is thus reached and is found to consist of the supposed Archean schists and huge intrusions of granite, which extend to the northern margin.

This zone of gneiss-granite, gneiss, amphibole schist, mica schist, phyllite, and crystalline limestones, intruded by massive coarse-grained granite, corresponds in strike with the Fu-niu-shan† (described in a preceding paragraph), and the petrographic characters of the rocks are similar to those of that range. The presumption is strong that the area is one of pre-Sinian schists and intrusives. They have not, however, been observed in unconformable relation beneath the Paleozoics, and their pre-Sinian age can not be considered established, in view of the fact that post-Paleozoic metamorphism and intrusion is general elsewhere in the region. Nevertheless, the weight of evidence is in favor of their antiquity, and if it be accepted we must regard the northeastern portion of the Ts'in-ling-shan as consisting of Proterozoic and Archean rocks. As we did not cross any western continuation of the ancient gneisses and granites in longitude 108° 30', the zone does not extend westward along the face of the range, as

---

* Reise des Grafen Széchenyi in Ostasien, vol. I, p. 448 *et seq.*
† *Ibid.*, vol. I, p. 446.

it was supposed to. It apparently runs to a point west of Si-an-fu, about longitude 109°, and is a triangular area bounded on the southwest by metamorphosed Paleozoics and granites intruded into them. The next section toward the west, across the Ts'in-ling-shan, is that observed by the expedition of 1904–05, between Chóu-chï-hién and Shï-ts'üan-hién, in longitude 108° 30′ E. It lies not more than 65 miles, 100 kilometers, west of Lóczy's eastern route and 80 miles, 125 kilometers, east of von Richthofen's.

In approaching the Ts'in-ling-shan from the Weï valley, in longitude 108° 15′, from Chóu-chï-hién, we expected to find the front of the range composed of granite, as shown by von Richthofen. Instead, the foothills and northern slopes consist of green schists, prevailingly chloritic, in which occur thin beds of quartzite and highly siliceous marble. These rocks dip toward the north at their northern margin, but a short distance south dip steeply southward. They maintain a uniform southern dip of bedding and schistosity for 5.5 miles to their southern limit at Liu-yüé-ho, where they are succeeded by white quartzite and massive gray limestone, folded in a syncline. The contact was not observed, but a coarse conglomerate of quartz, quartzite, and granite pebbles in a reddish or dark purple matrix is supposed to be the lowest stratum overlying the schists. It occurred in large masses in a brook, which flowed along and near the contact. These occurrences very closely resemble the relations observed by von Richthofen, who describes similar rocks in the statement quoted below. We agree with him entirely as to the petrographic likeness between the green schists of these sections and those of the Wu-t'ai system of Shan-si. Their position beneath a distinct series, which can not be younger than Paleozoic and from which they appear to be separated by a marked unconformity, is that of pre-Sinian terranes.

Von Richthofen's description of the rocks which he assigns provisionally to pre-Sinian series, can not be better stated than in his own words.* After describing the general aspect of the northern slope of the Ts'in-ling-shan as it rises from the plain of the Weï, he says, with reference to the Archean zone:

The first rock seen in place near Yi-mönn is granite, in part of medium grain, in part very coarse, a mixture of red orthoclase, some white plagioclase, quartz, and black mica. Here and there occurs a thinly laminated mica gneiss, which occasionally goes over into hornblende and chlorite gneiss. It is everywhere penetrated by granite. I saw no other rocks during this day. Typical gneiss and typical eruptive granite (with the exception of gneiss-granite) predominate probably in the entire width of the northern zone of the mountain range, for in the alluvial cones along the northern base, west of Kwei-tschönn, I saw only these rocks, which were there developed in great variety.

---

*China, vol. II, pp. 563 and 565

He describes the probable Wu-t'ai rocks thus:

South of Twi-tsze-shang, gneiss and granite give way to a series of rocks which are characterized by green color, and when the component minerals are macroscopically recognizable consist of abundant hornblende or chlorite, or of both minerals, and exhibit great petrographic variety. Predominant are rocks composed of reddish feldspar, blackish green hornblende, and dark scales of chlorite. The last often occurs in irregular, sometimes sharply fringed spots, with hornblende, which lie in a ground-mass composed essentially of feldspar. The chlorite then occurs next to the hornblende in such manner that one is inclined to consider it an alteration product of the latter. Dikes of granite, pegmatite, and quartz are common, especially in the northern zone. With reference to the arrangement of the constituents, as well as with regard to their distinct separation, the rocks vary greatly. Occasionally they are massive without recognizable parallel structure, and they then have in some places the character of serpentine, the feldspar being subordinate. More often the magnesium minerals are arranged in parallel surfaces, sometimes only partially, as is the case with mica in gneiss-granite; otherwise, however, so regularly that more or less nearly perfect hornblende and chlorite schists are produced.

With the first steps on this formation I was reminded of the Wu-t'ai schists of northern Shan-si. Shortly there followed the typical members of that formation, as a very thick sequence of green, partially schistose rocks, in which none of the constituent minerals are macroscopically recognizable. The entire system of schists has, as a whole as well as in its members, the constant strike of west 12° north, east 12° south. The dip is uniformly south, mostly at an angle of 50° to 70°, occasionally less steep. * * *

Before reaching Tsau-liang-yi other rocks are seen in great numbers in float from the eastern ravines. Predominant among them are dense, quartzitic rocks, whose fracture planes gleam with innumerable little feldspar crystals. These are probably altered clayey sandstone. Therewith occur conglomerates of rounded quartz, in a matrix which partly corresponds to the above-described rock, is partly purely quartzitic, and has a reddish to dark-violet color. As I was unable to find these rocks in place I can not say whether they correspond to the sandstones and conglomerates which occur in the lower part of the Wu-t'ai formation south of the temples of Wu-t'ai-shan, or whether they belong to the coal-bearing strata which are presently to be described.

It will be observed that the pre-Sinian rocks are more extensively developed in the western section seen by von Richthofen than in the one observed by our expedition. We did not find the zone which is described by him as Archean gneiss and granite, and the belt of Wu-t'ai schists is but 5.5 miles, 9 kilometers, wide, whereas that in his section is 20 miles, 32 kilometers, across. There is, indeed, little reason to suppose that the two occurrences are one and the same zone. The strike in von Richthofen's section, east 12° south, if extended, carries the belt which he saw at least 12.5 miles, 20 kilometers, south of that which we observed. This direction appears to be the trend of axes in the range, and it is probable that there are subparallel belts on distinct folds. If so, the eastern zone, extended northwestward, lies beneath the Weï valley, and the western, extended southeastward, ends before it reaches our route. We crossed an anticline in Paleozoic strata about in its line of strike.

Lóczy, whose observations of the eastern Ts'in-ling-shan have already been described, recrossed the range on the borders of Tibet, 65 miles, 100 kilometers, west of von Richthofen's route. Obrutchov followed the same way southward from "Hoj-shien" to "Quan-jüon-shien," and returned northward by a route that diverges to the west and leads over Min-chóu to Lan-chóu-fu. The two geologists agree in describing the structure as that of close folding, involving overturned folds and overthrusts. They both observed more or less metamorphosed Paleozoic strata, which make up the great mass of the range, and among which Silurian, Devonian, and Carboniferous horizons are distinguished by fossils.* They also noted metamorphic schists of uncertain age, which are doubtfully assigned to the "Archean" by Lóczy, and by Obrutchov are distinguished from the Paleozoies under the general term of "metamorphic schists."

The section in which the "Archean?" rocks are most prominent lies between "Lo-jan-shien" (Lo-yan-hién) and "Tschau-tjen" (Chau-tién) in longitude 106° 20', between latitude 32° 30' and 33° 30' north. It is in the strike of a wide belt observed by von Richthofen, 45 miles, 70 kilometers, further east, which on the basis of the general stratigraphic sequence was classed by him as of Paleozoic age; and 80 miles, 125 kilometers, still further east, near Ssï-móu-ti, we crossed gneisses, schist, and marble, which are intruded by large granite masses and which we also regard as metamorphosed Paleozoics.†

In the particular section of the western Ts'in-ling-shan under discussion, between Lo-yan-hién and Chau-tién, Lóczy mentions the occurrence of crystalline schists, gneiss, and phyllite, and crystalline limestone, which lie in isoclinal folds and are intruded by large bodies of diorite. He distinguishes argillites, chloritic schists, and epidotic schists, as well as partly metamorphosed Paleozoic strata, from underlying gneisses, with which they are, however, folded. And he cites occurrences of fossiliferous Devonian and Carboniferous strata in nearby localities. The sections which he has drawn show‡ that the rocks which, on account of petrographic likeness, he has classed together, occupy a variety of positions with reference to those whose age is more surely established. The schists under discussion are indicated by a red tint and blue dashes. In section 2, profile plate 2, north of Lo-yan-hién, they appear overlying less highly metamorphosed "Wu-t'ai?" schists on one side and "early Paleozoic limestone" on the other, in a syncline; south of Lo-yan-hién, in the "Pej-ho-scha," they underlie "Silurian" strata and rest upon granite. Still further south, beyond

---

* Reise des Grafen Széchenyi in Ostasien, vol. I, pp. 422–439, plates II and VI. La Face de la Terre, Suess, vol. III, pp. 270–271, fig. 35.

† Vol. I, pp. 308–310.

‡ Reise des Grafen Széchenyi, vol. I, pp. 465 and 428, plate II.

the broad zone of granite near "Jam-pa-quan," they appear to be intruded by granite dikes, and are folded in such a way as to constitute anticlines between which less highly metamorphosed strata, assigned to the Paleozoic, appear in synclines.

From the observations of Obrutchov, Lóczy, von Richthofen, and ourselves, in four parallel sections across the central and western Ts'in-ling-shan, it thus becomes apparent that the geologic structure involves several complex factors. There are metamorphic rocks, some of which are probably pre-Cambrian, and among them may be representatives of the basement gneisses as well as of the Proterozoic metamorphosed series. There are other metamorphic rocks of Paleozoic and probably also of early Mesozoic age, which vary in degree of metamorphism according to the intensity of action incident to local folding and intrusion. And there are areas of Paleozoic strata which exhibit little or no effect of metamorphism. These various masses are closely folded, overturned, and overthrust, producing relations as complex as some of Alpine structure.

We may conclude that the occurrence of pre-Sinian rocks in the western Ts'in-ling-shan is highly probable. Where such rocks are known in the central Ts'in-ling-shan, they are chloritic schists with thin beds of quartzite and marble, very closely resembling the strata of the Si-t'ai group of the typical Wu-t'ai district.

ANCIENT METAMORPHIC ROCKS BEYOND CHINA.

Under this heading I propose to consider the relations of certain rocks which are variously classified as Archean, Wu-t'ai, Sinian, early Paleozoic, and Silurian, by observers in Tibet, Indo-China, and India.

In Tibet they constitute the unfossiliferous pre-Devonian series of the great Nan-shan mountain system, and have been observed by Lóczy in his journey along the northern base of the Nan-shan range, from Lan-chóu-fu to Sù-chóu and return; and by Obrutchov in his far more extended journeys across southern Mongolia and back and forth among the several great ranges of the Nan-shan.

Under the heading crystalline schists,* Lóczy says:

The Archean complex of gneisses, mica schists, and phyllites plays a subordinate rôle in those parts of the middle Kuen-lung chain which we traversed. In none of the mountains which we crossed was I able to distinguish a crystalline axis.

He then discusses at some length various occurrences of crystalline schists, having in mind chiefly their eccentric position in the individual ranges in which they occur, but gives little information concerning their

---

* Reise des Grafen Széchenyi, vol. 1, p. 642.

geological relations. As constituent rocks of the Archean he repeatedly mentions gneiss, mica schists, amphibole schists, crystalline limestone, mica phyllite, gneiss-granite, and biotite-muscovite granite. Referring apparently to the same series of ancient crystalline rocks, Obrutchov frequently employs the terms "metamorphic sandstones and schists." With reference to the Barun Ula range,* he says:

The axial part of the range is chiefly composed of red and green Archean gneisses dipping steeply inward on both slopes, that is, forming a remnant of an ancient syncline.

In various sections of the different constituent ranges of the Nan-shan system which he crossed, for example in that of the Potanin range,† he distinguishes a group composed of gneisses, quartzites, micaceous schists, and intrusive granite, from semicrystalline schists, quartzites, and limestones. Describing a section across the Richthofen range, he says:

On the southern slope we see a thick series of supra-Carboniferous deposits, forming one of the southern ridges, the divide of the range. They strike north-northwest diagonally across the range and dip steeply inward on both sides. Below the peak Yang-kou-er appear more ancient formations, namely gray sandstones and shales, slightly metamorphosed, which I consider early Paleozoic. They do not closely resemble the ordinary metamorphic sandstone and schists of the Nan-shan and other parts of central Asia where I have seen them. Perhaps they are Silurian, perhaps still older.

It is much to be regretted that Obrutchov's third volume, in which he proposed to interpret the notes published in the two now available, has not appeared. As they stand the voluminous accounts of his observations offer little more than detached petrographic descriptions, from which geologic relations can hardly be deduced.

Lóczy distinguishes a system which he describes under the name Nan-shan sandstone, and which, after having studied von Richthofen's second volume, he referred provisionally to the Wu-t'ai and Sinian systems, but he did so only with the reservation that the correlation is suggested, but not established, by the likeness in lithologic character and degree of metamorphism exhibited by the rocks of the several systems. Regarding the Nan-shan sandstone, he states that it consists chiefly of gray-green sandstones and clay slates, which are frequently traversed by distinct cleavage or schistosity and which are barren of fossils, except for occasional indistinct impressions, that may possibly be ascribed to fucoids. From the various sections and descriptions given in his chapter on the northern slope of the Nan-shan range,‡ it appears that the Nan-shan sandstone is very intimately folded and is intruded by large masses of granite.

---

* Central Asia, North China, and the Nan-shan. Obrutchov, vol. II, p. 79 (in Russian).
† Ibid., vol. II, p. 115.
‡ Reise des Grafen Széchenyi, vol. I, pp. 532–559.

Another series described by Lóczy, which belongs in this category, is that which he designates early Paleozoic limestones. He says:*

At several points in the range of the Nan-shan I found light-colored, dense, half-crystalline, sometimes siliceous, dolomitic limestones. Everywhere the limestones occur in moderately thick strata, conformably bedded with the Nan-shan sandstone, either inter-bedded or in the form of "Klippen." In none of these limestones could I find fossils, even in numerous thin sections. The investigations of Dr. Konrad Schwager of the material sent to him likewise yielded a negative result. Lines which might suggest bivalves, a spheroidal inclusion with a tangled, reticulated character, and a globulitic structure, occasionally suggest an organic origin.

At the time that Lóczy and Obrutchov wrote their descriptions they could compare the formations which they had observed only with the Archean, the Wu-t'ai or Huronian, and the Sinian. According to von Richthofen's definition, the latter comprised strata conformably underlying the Cambrian, which might therefore be considered the most ancient Paleozoic deposits. In distinguishing the Nan-k'ou or Hu-t'o system as a pre-Cambrian series, separated from the fossiliferous beds by an unconformity and characterized by highly siliceous limestones and slight metamorphism, we have found a series which closely resembles the rocks which Lóczy designates early Paleozoic. As the limestones to which he thus refers are interbedded with the Nan-shan sandstone, some part of that system is also probably to be referred to the Nan-k'ou (Hu-t'o) terrane.

Among the more highly altered metamorphic schists, quartzites, and crystalline limestones, there may probably be found the representatives of the Wu-t'ai system. It would not be surprising if outcrops of these ancient formations should be found to lie in one or more rudely concentric arcs corresponding to the strike of the mountain ranges, which extend from the type locality, the Wu-t'ai-shan in northern Shan-si, through southern Shan-si and across Shen-si and Kan-su to the Nan-shan in Tibet. This is the outline of the continental region during the succeeding Sinian age, and it would be in accord with the general laws of relation between continental structure and mountain trends that representatives of the Wu-t'ai system should be found at intervals along this arc.

In his southward journey from the northern Tibetan regions where the Nan-shan sandstone is typically developed, through western Yün-nan to Burmah, Lóczy observed formations which he correlates with it.† They occur chiefly between Ta-tsién-lu and Ba-t'ang in the outer ranges of the

---

*Reise des Grafen Széchenyi, vol. 1, p. 651.
† Ibid., vol. 1, p. 724.

Alps of eastern Tibet. He describes the rocks as monotonous clayey sandstone, of gray and dark colors, associated with clay schists and also with amphibolite and chlorite schists. The presence of the latter leads him to correlate the series with the Wu-t'ai. Semicrystalline limestones occur beneath the sandstones, the strata being extensively intruded by granite.

Throughout much of Indo-China there occur ancient schists which are described by Fuchs and Saladin,* who distinguish two varieties, the one greenish, siliceous, hard, and compact, apparently the upper part of the terrane; and the other a gray, lustrous schist. These metamorphic rocks are associated with granite, which, as we may infer from the language of the authors, is probably intrusive. The petrographic description is sufficiently close to that of Lóczy to suggest that these metamorphic strata correspond with the supposed Nan-shan sandstone occurring between the Ta-tsién-lu and Ba-t'ang, especially as the mountain ranges and structural features are continuous. The schists are, however, classed by Fuchs and Saladin as possibly Silurian (probably in the sense of early Paleozoic) as they underlie fossiliferous Devonian schists and sandstones; but the stratigraphic relations have not been worked out and it is not clear how old the pre-Devonian rocks may be. In a recent letter from M. Emm. de Margerie my attention is called to the occurrence of extensive and massive conglomerates which may represent the lower Cambrian or pre-Cambrian glacial deposits in this region.

Westward from Yün-nan and Burmah in the Himalayas occur ancient rocks which were described as an older granite-gneiss and a younger series of schists (the Vaikrita system†). Hayden‡ states that the granite-gneiss is intrusive in Cambrian and Permian strata, and that certain schists assigned to the Vaikrita system are altered Cambrian slates and quartzites. In the Peninsula region of India gneisses have been distinguished as older and younger, but without sufficient basis in observed relations, according to Oldham.§ No distinctly sedimentary pre-Cambrian rocks are known there.

Thus we are not as yet able definitely to distinguish metamorphic Proterozoic rocks in south China or India in such a way as to draw a parallel with the Wu-t'ai schists.

---

* Explorations des Gîtes de Combustibles de l'Indo-Chine, Annales des Mines, 8 sér., Mém. 2, p. 205.
† Griesbach: Geology of the Central Himalayas, India Geological Survey Memoirs, xxiii, p. 40.
‡ Geology of Spiti, India Geological Survey Memoirs, xxxvi, pt. i, p. 8.
§ Manual of Geology of India, page 23.

PRE-SINIAN DIASTROPHISM.*

Diastrophism is a phenomenon which finds expression in the oldest rocks as well as in the youngest mountains of all continents, and of this general fact Asia is the most striking illustration which the earth presents. Large areas exhibit rocks which have been intensely deformed and are among the most ancient known; and the greatest mountain chains challenge credulity by the evidence of their extreme youth. The earlier diastrophic movements escape us in the mists of unrecorded ages of earth-history; we take up the observation only where the facts become partly intelligible; but from that remote time to the present we find a connected series of events.

Suess describes these facts and draws the conclusion that:

The compressive force formerly acted throughout the entire expanse of the globe, whereas now it is localized in certain special regions.†

An alternative view may be stated: namely, the widely distributed effects of schistosity are produced in masses that lie in a relatively deep-seated zone, where movement is somewhat uniformly distributed; whereas, special regions of folding are peculiar to a relatively superficial zone, in which movement is localized and concentrated by special conditions of structure and resistance. That is to say, the distinction between widely distributed and localized structures is one of depths rather than of times. This view is stated more in detail in the final chapter of this volume. Applying it to the explanation of the general structure of the Archean, we may reason that the masses which that structure characterizes were deformed in a zone at such a depth that they were below any localized orogenic effects. That there were regions of special plication, even in the Archean is probable, but the superficial folded masses have been eroded.

Gneisses and schists constitute the T'ai-shan complex of China and also the corresponding rocks described by many observers in Siberia, central Asia, India, and Indo-China, and designated "older Archean." They are everywhere fundamental rocks; their constitution is very complex; they consist of minerals resulting from extreme metamorphism under great pressure; and their larger structure exhibits repeated intrusions, often

---

*Following Powell and Gilbert (U. S. G S. Monograph 1, pp. 3-340) I shall use the term *diastrophism* to denote all the processes of deformation of the earth's crust, and shall distinguish between *orogenic* movements, which result in the commonly observed phenomena of mountain ranges and mountain structure, and *epeirogenic* movements, which are expressed in the elevation or depression of broad areas and are recorded in eroded surfaces or accumulated sediments. The distinction is one which has found little, if any, recognition in the literature relating to Asia, but which is fundamental and clearly recognizable in the geologic facts.

†La Face de la Terre, vol. III, p. 7.

along parallel planes. · It is commonly recognized that they acquired these characters when deeply buried beneath masses, which subsequently were eroded in consequence of exposure above sea-level.

· To pursue this thought would lead too far afield in the direction of theory. Adhering to the facts, we may consider the leitlinien, or axial trends, of the Archean in Asia as developed by Suess.* They maintain two principal courses, the one known as the Saian direction striking east-southeast or southeast; and the other, the Baikal direction, east-northeast. Trends having these bearings meet south of Irkutsk, where they were first distinguished by Tchersky,† and there inclose on the south the amphitheater of Irkutsk. By assembling the geologic observations for all northern Asia, Suess has traced these axial trends and shown that they characterize an immense area of eastern and western Siberia. The peripheral ranges extend from the island of Sakhalin across the Amur into northern Mongolia, in the Baikal trend, and thence in the Saian direction through the Saian mountains to the Ob. This structure is not only of vast extent; it is also fundamental, as well through its antiquity as through its nucleal position. About it are arranged the other structures of Asia. They are pendant from it, as Suess happily says, like garlands.

Succeeding the Archean, in the sense given that term in these volumes, the next oldest rocks of which we have definite knowledge are the Wu-t'ai schists of northern Shan-si. They are early Proterozoic. They probably are not younger than many terranes classed as Archean in the broader sense of that term commonly employed by writers on Asiatic geology, for similar strata have not, as a rule, been sharply distinguished from the basal gneisses.

The Wu-t'ai schists are distinct sediments, which represent familiar conditions of erosion in a sequence that is generally recognized among later deposits. Unfortunately our interpretation is in a measure balked by uncertainty of the exact stratigraphic succession and ignorance of the former extent and distribution; but I proceed on the basis of our present understanding.

According to that, the lower Wu-t'ai strata are heterogeneous, quartz-ose and clayey in the lower part, more argillaceous in the middle, calcareous above; and this lower series is unconformably succeeded by a great thickness of shale which carries a basal conglomerate of large quartz and quartzite pebbles. Regarded as products of erosion the strata have a perfectly normal sequence, which corresponds to a complete erosion cycle

---

*La Face de la Terre, vol. III, chapter III, Le Faite Primitif. Summary on page 138.
†I. D. Tchersky. Sur la tectonique des montagnes de la Sibérie Orientale. Trav. Soc. des Naturalistes, St. Petersburg, XVII, No. 2, 1886. Procès-verb. pp. 52–58.

and subsequent prolonged continental phase, or marine transgression. The latter is more probable, but we can not on present evidence exclude the possibility that the upper Wu-t'ai strata accumulated subaerially on a low-lying land. Confining our attention for the moment to the lower series, which ranges from quartzose argillaceous to calcareous, we recognize a succession of rocks which, though much older, resemble the Sinian (Cambro-Ordovician) terrane. At the base are the weathered, oxidized products of subaerial rock decay, such as, in consequence of exposure above the plane of gradation, are transported by streams and delivered to the sea. Toward the top are the finer, finally chiefly calcareous sediments, which result from the topographic and climatic conditions that follow from prolonged fixed relations of land and sea. The unconformity indicates an interruption of those relations which, if the overlap of the next later strata on the Archean be as great as we suppose, was a. very notable interruption. The upper series represents ultimately renewed erosion, transportation, and deposition. The Wu-t'ai deposits are several thousand feet thick. They are at least equal to the Cambro-Ordovician in duration of time and altogether may equal most of the Paleozoic. The nature and volume and sequence of sediments correspond with those which represent epeirogenic movements and static conditions of later periods; they evidently indicate equivalent movements and conditions for those early times.

The Wu-t'ai strata were folded and intruded by igneous rocks, and the whole mass was rendered schistose before the next succeeding formation was laid down in the typical district. The igneous rocks became gneisses; the heterogeneous strata became biotite, muscovite, and chlorite schists; and the limestones changed to marbles, with the development of garnet, staurolite, and other minerals from the shaly beds. Metamorphism was preceded by or associated with folding and thrusting into isoclinal structures. The phenomena compare in intensity and volume of rocks affected with the effects shown by the Paleozoics of the Han region, China, or of the Sierra Nevada, California, as a result of deformation during the Fermo-Mesozoic.

Thus we see that the Wu-t'ai region passed through an epoch of intense orogenic activity at the close of the Wu-t'ai period of the Proterozoic era.

In other districts than that of the Wu-t'ai-shan, Shan-si, where strata are identified by lithologic character as being probably of Wu-t'ai age, the structural facts are less well known. In the Ts'in-ling-shan, Shen-si, chloritic schists with some limestone and quartzite occur unconformably beneath Paleozoic strata, from which they are markedly distinguished by greater metamorphism. The suggested events of deposition and orogenic disturbance are comparable in time and character with those of the typical region.

Regarding the great Kuen-lung system of Central Asia, the available data clearly distinguish highly metamorphosed sediments beneath Devonian or pre-Devonian strata. They appear to fall into two groups, of which one consists of schists like the Wu-t'ai schists, rocks of the other group being less altered; and the facts indicate a sequence of activities which resemble those of Wu-t'ai time.

Thus in an arc which extends from the Wu-t'ai-shan in Shan-si around the southeastern and southern margins of the Mongolian plateau, there may be traced evidence of a very ancient mountain movement or group of movements. The axial trends of the early structures correspond in a degree, though not exactly, with the existing ranges, and this coincidence forms a reasonable basis for the view that certain conditions, possibly mechanical, have controlled and still control the courses of mountain chains. The coincidence does not demonstrate the continued existence of the Wu-t'ai or Kuen-lung ranges as mountainous elevations from early Proterozoic time to the present; in view of the stratigraphic evidence that Sinian lands were low, or of the physiographic evidence that the present mountains have grown up largely since the Tertiary, such a hypothesis is untenable. Nevertheless, it is one which was commonly accepted thirty or forty years ago. Von Richthofen regarded the Kuen-lung line as a great divide established in a very early period,[*] and describes the Ho-shan[†] as a monument of the oldest time, which formed a long island in the Carboniferous sea. King's views of the antiquity of mountain heights of the United States are clearly expressed in the volume on the systematic geology of the Fortieth Parallel. The effectiveness of erosion and the significance of strata as records of highland or lowland conditions have since received more adequate recognition. Beginning with Powell's conception of a base-level and the demonstration that the Appalachian folds had been planed,[‡] the generalization has been developed that existing heights are nearly all post-Cretaceous and largely post-Miocene. But at the same time the view has been strengthened that zones of mountain growth are so conditioned that repeated elevations occur on the same sites with similar trends, though at widely separated intervals. In the latter sense the Wu-t'ai and Kuen-lung axes of mountain growth are very ancient, one may say primeval, features of Asia.

Traces of the mid-Proterozoic epoch of orogeny are found in Liau-tung and Shan-tung, the eastern mountain provinces, in meager occurrences of pre-Sinian sediments, which are so metamorphosed as to be compared to the

---

*China, vol. II, pp. 647–648 and 709.
†*Ibid.*, p. 457.
‡W. M. Davis, Rivers and Valleys of Pennsylvania, Nat. Geog. Mag., vol. I, p. 183, 1889.

Wu-t'ai schists. Von Richthofen enumerates them* and nothing has been added to his account except Blackwelder's notes on the Ta-ku-shan quartzites. The facts do not suffice to fix any dates other than that one which preceded late Proterozoic. In the great lapse of earlier time there were in North America, and probably also in Asia, several cycles of erosion, sedimentation, and orogeny, and the correlation on lithologic likeness is inconclusive. We believe the Wu-t'ai terranes to be divided by an unconformity, and similar divisions probably exist elsewhere. Hence we can not establish equivalency of the epoch of folding in Shan-tung with that which was so decided in the Wu-t'ai, but in view of the intensity of effects in both regions, a rough correlation has a presumption in its favor.

The orogenic activity, which affected the Wu-t'ai rocks before any later beds known to us accumulated, divides the Proterozoic of China into two major periods. The earlier, the Wu-t'ai, links itself to the much older Archean; the later, the Hu-t'o or Nan-k'ou period, is like the early Paleozoic in general character. Whatever elevations resulted during the dividing epoch of orogenic activity, they did not survive to deliver sediments to the Nan-k'ou seas. Except that they may be represented by rocks which are not yet separated from the late Proterozoic, we must suppose that the mid-Proterozoics were deposited beyond the confines of the continent or in the deep synclinoria which in pre-Tertiary time traversed it.

Strata of the Hu-t'o system in the typical district succeed the Wu-t'ai after an interval represented by the profound orogenic activity described in the preceding paragraph. The sequence of sediments ranges from clastics of rather fine grain to carbonates; quartzites interbedded with greater thicknesses of slates pass by transition upward into cherty limestones. The total thickness is very roughly guessed at 5,000 feet, 1,500 meters. The clastics are terrigenous deposits, probably of the littoral zone. The equivalent marine formations are the lower limestones of the Nan-k'ou formation, of which the upper part is probably identical with the limestone of the Hu-t'o system. The unconformity and succeeding strata mark a transgression upon a flat land, which was sufficiently warped to deliver to the coastal waters the sands and clays of residual and alluvial deposits. Off shore, at a distance of 50 miles, 80 kilometers, or less, limestone formed from the very inception of the transgression, and extended landward as the sea widened. Such is the record for North China. If we rightly correlate the Nan-shan sandstone, or that part of it which carries the "early Paleozoic" limestones of Lóczy, with the Hu-t'o system, the evidence of similar shore conditions may be traced westward across Tibet.

---

*China, vol. II, p. 707.

Late Proterozoic strata are not known in Shan-tung, which seems to have formed an eastern shore of the strait in which the Nan-k'ou limestone was deposited. Nor have they been noted elsewhere in Asia outside of Tibet, unless some of the rocks of western Yün-nan, which Lóczy assigned to the Nan-shan terrane, prove to be of that age.

In their typical occurrence in northern Shan-si, Hu-t'o strata are discordantly overlain by the earliest Sinian (Cambrian) sediments. The older rocks were evidently folded and eroded to a peneplain, over which the Sinian sea transgressed. Where we observed the Sinian in relation to the Nan-k'ou limestone (at Nan-t'ang-meï, Chï-li) an unconformity is indicated by a basal conglomerate of chert.* A discordance of dip was not observed, but probably exists; yet the strata may be parallel. There are no other observations which enable us to determine the area affected by the post-Nan-k'ou pre-Sinian disturbance. The effects may be looked for throughout that littoral zone in which the type locality lies and which probably extends along the Kuen-lung system. The movement did not compare in intensity with that which closed the Wu-t'ai period; and even though it should ultimately be found to have affected an extensive belt, we can not assign to the epoch of deformation and consequent erosion a duration at all equivalent to that of mid-Proterozoic diastrophism. Nevertheless, it appears to mark a break in continental history equivalent to that at the base of the Cambrian in North America, and consequently to be properly regarded as the last episode of the pre-Cambrian, immediately antedating the Sinian transgression and the advent of that Lower Cambrian fauna which is abundantly preserved in the Sinian deposits. The limitation of the term Sinian to a Cambro-Ordovician system is discussed in the next chapter.

Our present knowledge of pre-Cambrian diastrophism in eastern Asia may be summed up as follows: During very early times movements which are expressed in schistosity and metamorphism of the most ancient rocks occurred in a deep-seated zone. The superficial effects of the compressive movements are lost; that is, the fractured and simply folded rocks of that early time were eroded and the schists were exposed. The wide-spread occurrence of the Archean schists shows that epeirogenic movements were general. Of localized orogenic phenomena we have no direct evidence, yet we can not doubt that they also developed in a commensurate scale. The axial trends of Archean rocks are north-northeast (the Baikal direction) and west-northwest (the Saian bearing). These meet south of the amphitheater of Irkutsk, and further south are connected by the trends which correspond with the Altai ranges of northern Mongolia.

*Vol. I of this report, p. 131.

During the early Proterozoic a synclinorium developed on the site of the present Wu-t'ai-shan of northern Shan-si, and in it were deposited sediments which correspond in sequence with the progress of a cycle of erosion from youth to old age and subsequent marine transgression. These sediments were probably common to a zone which extends from the type locality northeast and southwest; in the latter direction it enters the Ts'in-ling-shan and thence curves west to northwest across Tibet in the Kuen-lung system. The corresponding diastrophic movements were epeirogenic, and they passed into a phase of quiescence such as has since repeatedly characterized such movements. It is probable that there was more than one epoch of erosion, at least one unconformity being recognized in the sedimentary series.

At a period which may be described as mid-Proterozoic, the zone of early Proterozoic sediments was sharply and intensely deformed. The disturbance was apparently accompanied by granitic intrusions of large volume, which were then or afterward rendered schistose in common with the folded strata. The events were complex. The movements may be classed as orogenic, since they resulted in deformation of strata by folding and thrusting in an apparently well-defined zone. The strikes follow the Baikal direction, northeast-southwest in North China, change in the Ts'in-ling-shan to east-west and northwest, and extend to the Kuen-lung. They thus form an arc outside that of the Baikal-Saian curve, embracing Mongolia, as was perceived by von Richthofen* and has been brought out by Suess† and others. This period of orogeny was for the provinces affected equally as important, apparently, as the intense orogenic disturbances which mark the Fermo-Mesozoic period in Central Asia and the western United States. It was very possibly an incident of a period of diastrophic activity such as closed the Paleozoic.

After a period of erosion, during which the altitudes that had resulted from the preceding activity were materially or completely reduced, but of which no sedimentary record is known, there followed a cycle characterized by the deposition of littoral sediments in the typical district of the Wu-t'ai-shan, and of marine limestones in a trough which traversed eastern China. Strata, which may be equivalent, occur in the Nan-shan range of northern Tibet. The deposits again correspond with the phases of an erosion cycle from youth to old age, and finally represent a transgression over a low continent.

The last event of the Proterozoic (pre-Sinian) era was a movement which is recorded in folding of the latest pre-Cambrian sediments; although possibly a local phenomenon of the littoral zone, so far as is yet known, it

*China, vol. II, pp. 635 *et seq.*, 647.
† Face de la Terre, vol. III.

occasions a decided unconformity of structure in the Wu-t'ai district and is represented by an erosion interval, even where there is no known discordance of dip with the Sinian. It is consequently regarded as an interval of sufficient importance to distinguish the Hu-t'o or Nan-k'ou system of the Proterozoic era from the Sinian system of the Paleozoic era.

PRE-SINIAN UNCONFORMITY.

The unconformity at the base of the Sinian system divides the Paleozoic from the pre-Cambrian. It is a break of the first magnitude, corresponding to a period of deformation and erosion, even where the underlying strata are the Ta-yang (Nan-k'ou) limestone of the late Proterozoic. More commonly the subjacent rocks are Archean and the hiatus embraces the much longer times which are in some localities represented by the two Proterozoic systems of strata and the sum of deformations and erosions they have undergone.

The latest cycle of erosion with which the Proterozoic closed was very complete. The land surface was reduced to a nearly perfect plain, upon which the transgressing sea of Sinian time found few eminences to level. In this statement we agree with von Richthofen as regards the completeness of planation, but differ from him in assigning to erosion the greater part of the work accomplished. In accordance with the prevailing views of the time in which he wrote his second volume of China, he therein held that the major work of denudation was due to marine abrasion. He says:

The first great occurrence after the episode of folding was an extensive abrasion, by far the most important which is to be demonstrated in the geological history of China. The complete difference between the tectonic movements which occurred before and after the period marked by this boundary suffices to show that a very long time interval lay between the happenings of the fifth and seventh phase, since a portion of the earth's crust which previously had been capable of intense internal movement was so stiffened that it thereafter only changed its level as a whole; but we obtain a more definite idea of the great length of this sixth period when we consider that enormous mountain masses which covered continental areas were completely swept away. From the fact that formations whose thickness can be measured only by tens of thousands of feet occur in single troughs, we are obliged to draw the conclusion that these are relatively small roots of former folded mountain chains, which attained at least the height of the Alps, but then almost completely vanished. We have repeatedly referred to the breakers of an advancing sea, which transgressed over the land in consequence of simultaneous mechanical erosion and positive subsidence, as the one agent which is able to produce a level surface where there previously existed an extensive mountain range. However much assistance may be given by atmospheric influences and the running water of the mainland, this alone is competent to produce an almost even surface of great extent. That force is the only one which we may here postulate. The surf removed not only the folded masses of the ancient formations, but also attacked their foundation, the old gneiss, and developed a plane of abrasion which extends across the remaining portions of the gneiss as well as over the synclines of the younger

Archean strata and the Korea granite, as is clearly proven by the uniform occurrence of the wide-spread covering of Sinian sediments. The surface was not, however, perfectly uniform, as the eroding force could not overcome the resistance of the hardest rock. The quartzites stand up in high reefs, as we have seen in Liau-tung at Sai-ma-ki and Ta-ku-shan. The Korea granite stands out occasionally in steep high bluffs, and yet in the immediate vicinity it occurs as the surface of a plain of abrasion on which Sinian strata lie in horizontal and undisturbed position. The irregularity of the action appears especially about the hard cores of gneiss in the mountains. We have seen how the layers of the principal mass of gneiss, which, around the Yellow Sea, strike north-northwest south-southeast and dip steeply, were involved in the thrust in the Sinian direction and thrown into disturbed positions which probably facilitated erosion, while near these deeply eroded and decomposed masses wild ridges of unweathered gneiss stand up. We must consider them the massive cores which were not destroyed by the surf. They form the characteristic mountains of eastern Shan-tung and Liau-tung.*

The view that notable elevations were cut away by the waves is not tenable in contradiction of the evidence of Sinian sediments. Where waves do attack bold coasts of gneiss, granite, schists, and quartzite, they spread coarse deposits of the obdurate rocks. Where large masses are rapidly leveled by subaerial erosion or marine abrasion, sediments of corresponding volume gather in some adjacent basin. The Sinian deposits do not represent either of these conditions. They are neither coarse nor voluminous. They are frequently fine-grained calcareous shale or thin-bedded limestone at the very base. The mechanical sediment of the basal formation has the character of a fine alluvium and is of uniformly moderate thickness, 350 to 500 feet, 105 to 150 meters. The material is red soil, particles of ferruginous clay being thoroughly oxidized and grains of sand coated with ferric oxide. The plane of contact at the base is sharply defined, usually very even, not broken by abrupt hollows or decided projections, but swelling gently over rounded bosses of the harder rocks. Pebbles of the subjacent rocks are wanting in the basal deposits, as a rule, and where they occur are limited to very local accumulations. Beds of arkose have not been seen, nor even beds of clean sand such as waves usually spread. Thus none of the effects of violent breakers are present; the evidence is that a gentler agent cleaned the surface of the ancient rocks. The facts support the view that the lowest strata of the Man-t'o formation were laid down in the shallows, lagoons, and flood-plains of a very low, flat coast, where weak waves, feeble shore currents, and rivers interacted.

The smooth surface of the old rocks is one which has been swept clean, as by the wash of little waves or by the wear of a stream engaged in lateral corrasion. It is probable that a layer of saprolite was removed in the process. The material laid down on the bared rocks is usually mud, occa-

---

*China, vol. II, pp. 710–711.

sionally mingled with coarser stuff. In the special case of the Yung-ning sandstone of Liau-tung, it is cross-bedded like a stream deposit from swift and variable currents; usually it is evenly stratified after the manner of deposits from quiet waters or on flood-plains. It is barren of fossils up to 100 feet, 30 meters, above the base, but above that horizon marine forms are evident.

Hence it is reasonable to infer that the zone of unconformity, comprising the immediately subjacent rocks, the contact, and the directly superjacent strata, represents a coastal plain reduced through erosion and lateral corrasion by streams to an even surface; covered during an early stage of subsidence relatively to sea-level by alluvium, and buried beneath fine, ill-assorted shore deposits of a shallow, rippling, advancing sea. Where the waves removed the alluvium the marine strata are among the lowest; where they did not, the bottom layers are of fluviatile origin.

It is possible that such a coastal plain should be diversified by an occasional hill of resistant rock, and hills of that kind may still survive if preserved through burial in sediment. Where the strata have in a recent geologic epoch been eroded from about them, they may again appear as prominent features of the landscape. Such was von Richthofen's view regarding the quartzite ridge of Ta-ku-shan and others like it. While granting the possibility of individual cases occurring, we do not agree to the statement of the final sentence in the preceding quotation: "They constitute the characteristic mountains of Shan-tung and Liau-tung." We consider the present relief of those peninsulas to be a relatively very modern phenomenon, and hold that resurrected hills of the early Sinian epoch are few. No case of the sort came under our observation.

The preceding statements apply to the unconformity at the base of the Sinian, wherever we saw it: in Shan-tung, Liau-tung, Shan-si, and Hu-peï. Its character as a plane is uniform over a stretch of a thousand miles, representing several hundred thousand square miles of contact. It is a feature of southeastern Asia, from latitude 30° to 42° north and longitude 108° to 124° east.

Near Ta-tung-fu in northern Shan-si, von Richthofen observed an occurrence of characteristic Sinian limestones, having at the base red, clayey, and calcareous thin-bedded deposits, which rest unconformably on gneiss. He remarks that the "Untersinisch," i. e., the Nan-k'ou limestones, are here wanting. The occurrence is one of overlap of the Cambrian strata, to which we restrict the term Sinian, beyond the pre-Cambrian, Nan-k'ou, limestones or their littoral equivalents onto a much older pre-Cambrian (Archean?) gneiss.*

---

*China, vol. II, p. 358.

A similar plane of unconformity presumably characterizes northern Siberia, where the great plateau of flat Paleozoic strata north of Irkutsk stretches from the Lena to the zone of folds adjacent to the Archean areas along the Jenisseï. The base of the Paleozoic is there Lower Cambrian, and the surface beneath it is of the same prolonged cycle as that in south-western Asia.

An unconformity comparable with that at the base of the Cambrian system occurs throughout Central Asia also, but the superjacent strata are probably Devonian or Silurian. The surface developed on the pre-Cambrian rocks is therefore younger than the Sinian, and there is room to question what features existed in Central Asia during the Sinian period. It is probable that the epicontinental sea did not spread over the entire region where the corresponding strata are now absent, although it no doubt covered some part. Whatever land area was exposed at any stage of the advance and retreat of the waters was then being eroded and furnished the sediment of the Sinian strata, which, had there been any considerable height of land, would consist of shaly and sandy deposits. They are, however, limestone, and it is a fair inference that practically all Asia, draining to the Cambro-Ordovician sea, was low and featureless.

The fact that Asia at the opening of the Paleozoic era was a featureless continent has important bearings. It limits the antiquity of mountain ranges, some of which have been discussed by eminent writers as of pre-Cambrian date, as elevations which have survived since that remote time; and it affords a basis of inference regarding a cycle of inactivity, which was common to other continents as well.

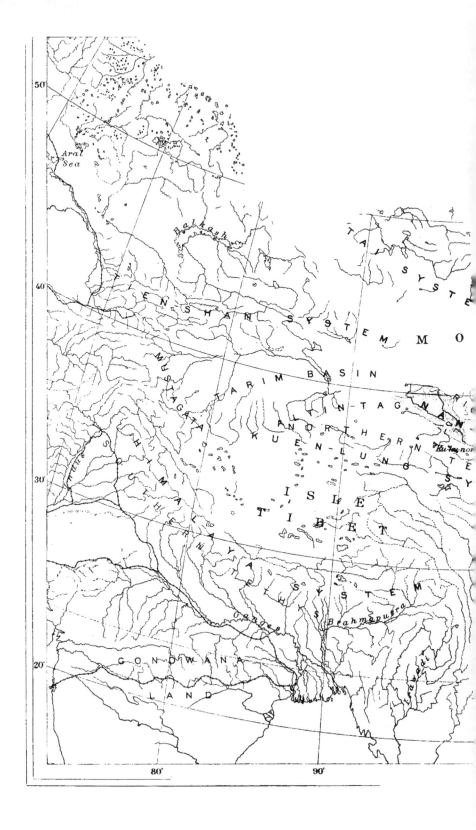

PLATE 3

120°   130°

LEGEND

UNDIFFERENTIATED
PROTEROZOIC STRATA
(Areas within which they
are known to occur)

UNDIFFERENTIATED
PROTEROZOIC STRATA
(Some areas within which
they are supposed to occur)

NAN-K'OU (HU-TO) STRATA
(Areas in which they
are known to occur)

WU-TAI SERIES
(Area of typical occurence)

CHAPTER III.—EARLY PALEOZOIC.

SINIAN SYSTEM (CAMBRO-ORDOVICIAN).

The name and its application.—Sinian was first applied by Pumpelly to the prevailing structural axes of eastern Asia, which trend northeast and southwest.* It was adopted by von Richthofen to designate a series of conformable strata which exhibit folds having the Sinian direction. They are characterized in part by Cambrian fossils, but were believed by him to extend downward below the base of the Cambrian, and at the top to include part of the Ordovician. The term is here used to designate the Cambrian and Ordovician strata to which he applied it, but those limestones which underlie the lowest fossiliferous Cambrian are excluded, after conference and agreement with von Richthofen himself.

The problem which confronts us in determining the base of the Sinian is inherent in the more or less local nature of an unconformity. Each unconformity is somewhere represented by continuous, conformable deposits, and the area of unconformity is bounded by areas of conformity. When we pass from one to the other there is difficulty in dividing the continuous series of strata at a plane corresponding to that indicated by the discontinuity in the neighboring series. This condition exists at the base of the Cambrian in certain localities in the United States, where the lowest fossiliferous Cambrian strata are conformably underlain by great thicknesses of sediments, that accumulated in the depressions from which the Cambro-Ordovician epicontinental sea expanded. Such sediments are by some regarded as pre-Cambrian, by some as the downward extension of the Cambrian. There is no difference of opinion regarding the base in sections where the unconformity intervenes, as is commonly the case.

In China there is usually an unconformity at the base of the distinctive red formation of the Sinian, and Cambrian fossils occur within 100 feet above the contact. Von Richthofen observed this conspicuous break, and we also obtained evidence of it in every section in which we saw the appropriate contact. But there are sections such as that of the Nan-k'ou pass. northwest of Peking, in which the unconformity was not noted by von Richthofen and may not exist. The strata there below the Cambrian are siliceous limestone, equivalent to that which is apparently unconformable beneath

---

* Smithsonian Contributions to Knowledge, vol. xv, Geological Researches in China, Mongolia, and Japan, p. 67. Sinian from Sinim, the name applied to China in the earliest mention made of that country—in Isaiah.

Sinian not far away, at Nan-t'ang-meï, and probably equivalent to the Hu-t'o system, which is separated from the Sinian by a decided break. Where it exists, the unconformity is everywhere taken as the dividing plane and the equivalent horizon must elsewhere determine the limits of Cambrian upon pre-Cambrian, even though the strata be locally conformable. Hence we are constrained to exclude from the Sinian certain limestones, which von Richthofen regarded as "Untersinisch," but which in all probability represent deposits that are pre-Sinian.

Thus used to name a series of conformable strata deposited during the great Cambro-Ordovician transgression of Asia, the term Sinian has wide correlative application, since a similar transgression spread over much of North America and Europe and was there accompanied by evolution of faunas closely related to those of Asia. From Cambrian to Ordovician there is general continuity of physical conditions and faunal evolution. A natural plane of division marked by unconformity frequently occurs near the middle Ordovician. The strata below that plane to the base of the Cambrian may be appropriately called Sinian.*

Von Richthofen gives the following description of the Sinian in his chapter on the rocks of Liau-tung:†

The above-described stratigraphic fact [a marked unconformity of dip] sharply distinguishes the formations which have been described [pre-Cambrian metamorphic rocks and intrusives] from a series of strata which we found to be widely distributed along our route. In regular succession they follow one another. They everywhere exhibit a richly varied stratification, which in the upper part is somewhat monotonous, but in the lower shows many peculiarities in consequence of the fact that the strata, which were deposited upon a sea bottom that was set with reefs, extend in some regions deeper than in others. At a certain horizon which is high in the series we found globulitic limestones that are especially distinct and afford a paleontological clue. According to a communication from Mr. Dames the trilobite fauna of Sai-ma-ki and the Tai-tsze valley consists essentially of the genera Dikelocephalus and Conocephalus, and is closely related to the fauna of the Potsdam sandstone of New York, Wisconsin, Iowa, and Minnesota, especially the province of the upper Missouri, and the relation is such that there is indeed no doubt of their equivalency. In the Fortieth Parallel survey the same fauna has been recognized by Hall in the mining districts of White Pine and Eureka. It seems that the Chinese fauna more closely resembles the American Potsdam fauna than the Swedish Cambrian. It has nothing in common with the Bohemian primordial fauna.

It is not appropriate to apply the name Cambrian immediately to the formation under discussion. For, until it may be possible to distinguish the paleontological horizon, we are obliged to assign to it an extraordinarily long sequence of strata, which constitutes a great whole in consequence of stratigraphic conformity, without any possibility of proving that the beginning and end correspond with those limits within which the name Cambrian is

---

* A. Geikie has translated the German term "Sinisch" as "Sinisian." The original English form used by the author of the name is "Sinian."

† China, vol. ii, pp. 107–108.

applied in Europe and America. It is, moreover, possible that the formation in China, in which strata containing the primordial fauna have indeed a definite but somewhat subordinate position, reaches much further down and therefore comprises a much longer period, while it also probably extends without noticeable interruption up into the lower Silurian.

SINIAN IN CHINA.

When asked to suggest a district in which the Sinian might be studied in typical character, von Richthofen referred us to Liau-tung and Shan-tung, the northeastern provinces of China. The conditions immediately preceding war rendered surveying in Liau-tung difficult at the time of our expedition, and Shan-tung became the scene of our detailed studies.

The Ch'ang-hia and Sin-t'ai districts which were selected for detailed topographic and geologic surveys are represented in plates XIII and XIV, volume I, and the local variations of strata are described by Blackwelder, volume I, chapter II. The general sequence consists of three well-defined divisions between two unconformities, as follows:

Unconformity by erosion.
Tsi-nan limestone (lower Ordovician).
Kiu-lung group, interbedded limestone and shale (Upper and Middle Cambrian).
Man-t'o shales (Middle and late Lower Cambrian).
Unconformity by dip.
Pre-Cambrian rocks.

Lower Sinian.—In Shan-tung, Liau-tung, and Shan-si, that is, throughout northern China, the characteristic strata of the Lower Sinian are red deposits, which we have called the Man-t'o formation. They are probably equivalent to von Richthofen's Tung-wön Schichten. In Central China, on the Yang-tzï-kiang, we saw nothing corresponding to the Man-t'o formation, the Sinian being composed of limestone, apparently to the very base.

The typical red shale of the Man-t'o formation passes into red or chocolate-brown shaly sandstone and is interbedded with thin but sometimes persistent layers of gray to cream-colored limestone. The thickness varies from 350 to 550 feet, 105 to 165 meters. The basal layers occasionally show local conglomerates, as at Nan-t'ang-meï, Shan-si, where a body of chert conglomerate rests upon cherty Proterozoic limestone, and south of Tung-yü, Shan-si, where the underlying slates of the Hu-t'o system constitute pebbles in the Man-t'o next the contact. Usually, however, the material adjacent to the eroded surface of pre-Cambrian rocks is highly oxidized residual soil or fine calcareous sediment, which is in either case foreign to the underlying metamorphic rocks.

The sandy red mud of the Man-t'o formation is an end-product of mechanical and chemical rock decay. Only the most enduring minerals, quartz, clay, and oxide of iron, remain in notable quantities. The material resulted from disintegration of crystalline rocks under climatic conditions favorable to oxidation, or passed through stages of accumulation as a continental deposit, during which it was subject to such conditions, and was laid down without undergoing chemical reduction or abrasion or sorting. The rock is red throughout, as the Permian and Triassic sandstones are, and like them was not only originally a red mud, but formed a red deposit. It did not become blue, as the red muds of Virginia now do in Chesapeake Bay in consequence of abundant organic matter. The persistence of red in the Man-t'o formation shows that organic substances were not present in quantity.

Waves, of which ripple marks on the sandy shales are here and there evidence, accomplished but little work in the way of sorting. They were evidently too weak to sort sediment in which the proportion of mud was so great.

Calcareous layers occur in the Man-t'o formation occasionally very near the base, persistently at horizons 100 to 150 feet, 30 to 45 meters, higher up, and again, less commonly, near the top, which is often sandy. A transition into the overlying limestone of the Kiu-lung group is formed of interbedded brown shale and gray limestone. Individual limestone strata are very uniform in thickness, though but a few inches, or at most 10 feet, thick and not of great extent. The more continuous are those near the middle of the formation, which were found in all the sections in the Ch'ang-hia and Sin-t'ai districts in Shan-tung.

The interbedding of the shale and limestone is irregular; it does not follow any rhythm. It is apparent that local conditions were unlike in adjacent waters at any one time and varied in unlike manner from time to time; but red sediment from the land or calcareous sediment from the sea was deposited at any time. The condition which favored precipitation of lime, whether chemically or organically, was continuously present and became effective whenever the environment became right; and so with the mud. It is not possible to assume that the limestones were laid down in deep waters; they are too closely related to the shales and sandstones which were deposited in shallows. As the limestones are relatively free from clay, and what they do contain is very fine, the lime-depositing waters were comparatively clear, and this clearness appears to have been the essential condition.

One may form a concept of the conditions somewhat as follows: Along the flat, red shore of the Man-t'o sea, bars and islands formed where streams emptied, and shut off the mud-carrying currents from intermediate stretches of coast. More or less extensive lagoons were thus produced and within these the waters were clear. Being partly closed and shallow, they were relatively warm and liable to maximum evaporation. Rippling of the surface favored precipitation of lime carbonate by agitation. Warmth and protection invited organic life, both plant and animal, which probably occupied the lagoons in low forms that did not become fossil before trilobites, the earliest that have been preserved, discovered the habitat.

The description of the Man-t'o formation has thus far dealt with it as it is developed in northern China. The red mud does not occur in the south on the Yang-tzï-kiang, where we saw the base of the Sinian, but the strata which we suppose to be equivalent are thin-bedded gray limestones which rest ön a well-defined glacial till. The latter was seen only near the village of Nan-t'ou, and we have named it the Nan-t'ou tillite.

Nan-t'ou tillite.—The Nan-t'ou glacial deposit occurs in longitude 111° east, latitude 31° north, about 200 feet, 60 meters, above sea. It evidently accumulated close to sea-level in early Sinian time, as it is overlain by marine limestones of that age. At the base the plane of the pre-Sinian unconformity is characteristically developed and covered by a cross-bedded quartzite, which may have been either river deposit or beach. The top of the quartzite is generally covered in the type locality and a cultivated slope interrupts the section for 100 feet, 30 meters. Above the terraced fields occur steep banks of tillite, a greenish rock, about as hard as unweathered shale, of irregular hackly fracture, not stratified, and containing pebbles and boulders of various kinds and sizes, many of which are striated. The thickness seen is 120 feet, 36 meters.

At the top of the tillite, beneath a cliff, is a well-exposed contact with the overlying limestone. The tillite passes into a greenish shale, consisting of the same materials, including characteristic pebbles, all rearranged by water. This shale conglomerate is about 2 feet thick and grades into the overlying limestone, the basal layer of a great thickness of Sinian.

The facts clearly demonstrate the presence at this spot of a glacier which gave way to marine waters and left a deposit of till that was slightly washed by waves before it was buried beneath calcareous mud.

The glacial deposit was seen only where it is exposed in the gorge of the Yang-tzï-kiang, beneath limestone cliffs. Both northward and southward from the river the escarpment stretches beyond sight, and continuous below it is the slope which, at the river, is formed of the tillite; but so common a topographic feature as a slope below a cliff affords little ground for inferring the extension of so unusual a deposit as an early Cambrian till, and we limit ourselves to the positive statement of occurrence at Nan-t'ou.

Whether the Nan-t'ou glacier was an exceptional occurrence or a representative of an extensive system, only in degree affects the deduction that the temperature of early Sinian time was low. Glaciation in latitude 31° near sea-level presents, it is true, a problem which refrigeration alone will not solve, especially as no traces of contemporaneous glaciers have been found further north; but there can be no doubt that it signifies severe cold throughout northern Asia. The fact agrees with the inference which may be drawn from the red sediments of the Man-t'o. That they were not reduced by organic matter proves the barrenness of the shores and seas, although life was abundant elsewhere before and during the Man-t'o epoch and soon after developed richly in the shallows; that it was at first absent and when it appeared in the sea was limited in variety may. be attributed to the low temperature.

Aridity was also probably a condition of the climate. Slight precipitation prevents glaciation, even under conditions of severe cold, as is the case in northern Siberia, and the absence of glaciers in the north in early Sinian time may thus be understood.

Middle Sinian, Kiu-lung group.—The Kiu-lung group of Shan-tung is a succession of limestones and shales which immediately follows the Man-t'o formation. Transition beds connect the two. Shale is a common rock in both, but in the Man-t'o it is red, whereas in the Kiu-lung it is green. Limestone is thin-bedded and subordinate in the former, in the latter it is usually massive and predominant. The Man-t'o contains a sparse Middle or Lower Cambrian fauna in its upper portion; the Kiu-lung carries very abundant faunas, which range from Middle Cambrian at the base to Upper Cambrian and possibly to lowest Ordovician at the top.

This Cambrian group was clearly recognized in Shan-tung, but not as distinctly elsewhere. In Shan-si it is represented by the lower part of the Ki-chóu limestone, which comprises all the Sinian except the Man-t'o shale; and on the Yang-tzï the Kiu-lung horizons are within the great limestone formation, which is the whole Sinian system.

I proceed to consider the Kiu-lung group as it occurs in Shan-tung. The type locality is the Kiu-lung range, a chain of hills which borders the Wön-ho valley on the south, southeast of the district town of Lai-wu-hién, longitude 117° 40′ E., latitude 36° 15′ N. The strata dip gently northward and are extensively exposed from the Man-t'o shale below to the Tsi-nan limestone above. The thickness is 900 to 1,000 feet, 275 to 335 meters.

The component strata are of green shale and limestone. The latter presents several distinct varieties: massive black to gray oolite, nodular or conglomeratic layers of shale, and uniform fine-grained blue limestone. These occur from the bottom up, in the order named, of such thickness and extent as to be considered distinct formations in some districts.

The black oolitic limestone, which commonly but irregularly occurs in the green shale above the red Man-t'o, reached a maximum thickness, as we observed it, of 550 feet, 165 meters, near the village of Ch'ang-hia,* and we called it the Ch'ang-hia oolite. In consequence of its massive character and vertical jointing it there forms imposing cliffs, and one does not suspect that it may, in a short distance, thin out to a layer of vanishing lenses. Such is, however, the case, as we found in the district north of Sin-t'ai-hién, where a stratum 100 feet, 30 meters, thick gave out in less than a mile. The color of the rock is due to the black oolites, with which it is usually crowded. Blackwelder has described their peculiarities,† and he finds that there is a series of forms ranging from true oolites, which exhibit a nucleus and concentric banding, through grades of finer to coarser crystalline texture, to a single crystal; and he concludes that all the bodies had a common origin as oolites, which formed concentrically around a nucleus. Some have remained but little altered and show the original structure, while others have more or less completely crystallized. Crystallization has progressed from a primary condition of many minute crystals toward an ultimate development of a few or one large crystal. The fossils of the Ch'ang-hia limestone are trilobites, brachiopods, etc., of Middle Cambrian age.

The middle portion of the Kiu-lung group is characterized by the predominance of shale. In the vicinity of Ch'ang-hia a single stratum 150 feet, 45 meters, thick was so clearly distinct that we described it as a separate formation, the Ku-shan shale. In the Kiu-lung hills and adjacent areas, however, no individual shale formation was distinguishable, the mass of shale and limestone layers being on the whole thicker, but irregular.

A special interest attaches to this member, since conglomeratic limestones, such as have been described by Walcott as "intraformational conglomerates," are of common occurrence in it. These peculiar rocks consist of an earthy calcareous matrix, in which flat, pebble-like bits of limestone are irregularly embedded. The pieces are commonly rounded, but sometimes sharply broken; they are fragments of thin limestone sheets, which were broken up, washed, and rearranged during the formation of the stratum in which they occur. Among the hypotheses that have been suggested to account for these conglomerates, we are restricted to those which regard the pebbles and the matrix as essentially contemporaneous. The fact of unbroken conformity with the immediately underlying stratum, which has been observed in every instance, excludes any hypothesis that presupposes unconformity and erosion of older rocks. Identity of earthy calcareous composition of pebble and matrix, and particularly identity of

* Vol. I, Plate XIII.
† Ibid., p. 30.

fossil content in both, confirms this. Characters common to shoal-water deposits mark these beds, which, though most frequent in the middle of the Kiu-lung group, occur also in the underlying Man-t'o. The conditions of occurrence, the constitution, and the detail of the conglomeratic layers require that during the deposition of calcareous mud in shallow waters there shall have been some layers that hardened more or less firmly to limestone strata. There is no evidence that they were extensive or continuous; they may probably have been limited and separate; but they were common. The mud containing these layers was disturbed and the more or less consolidated lime rock was broken, washed, and redeposited, after the manner of a conglomerate. A portion of the process as yet eludes interpretation. We do not know the physico-chemical or organic conditions under which limestones consolidate, and are therefore at a loss to understand why some layers or nodules harden before others. Our speculations are given in volume I, part II.

The upper part of the Kiu-lung group is a thick-bedded uniform limestone, of light-blue to gray color and usually smooth texture. From its development near the village of Ch'au-mi-tién in the Ch'ang-hia district, we called it the Ch'au-mi-tién limestone. Its thickness is about 580 feet, 175 meters, in the type locality, where it directly overlies the Ku-shan shale. In Kiu-lung hills the rock occupies a similar stratigraphic horizon and carries the Upper Cambrian fossils, which characterize it in the type locality, but its limits are not so clearly defined. In marked contrast to other Sinian strata below it, the Ch'au-mi-tién limestone is horizontally continuous and uniform. It represents a wide-spread condition of deposition, such as the circulation of a broad marine current over a continental shelf, and thus differs from the strictly littoral aspects of the Man-t'o terrane, and the inconstant phases of the shaly part of the Kiu-lung division.

Upper Sinian.—The Tsi-nan limestone, so named after the capital city near which it is exposed, is the highest formation of the Sinian system in Shan-tung. It differs from the underlying Ch'au-mi-tién, being less plainly stratified, dark gray to brownish in color, and poor in fossils, which are of lower Ordovician types. It may commonly be divided into two members: a lower, 250 feet, 75 meters, thick, consisting of shale and coarse crystalline dolomite, which weathers like calcareous sandstone; and an upper, 2,500 feet, 750 meters, or more, which is dolomitic limestone. The total thickness is indeterminate, since the upper surface is one of erosion, even where it is covered by later sediments.

This Ordovician limestone was confused by von Richthofen, who found no fossils in it, with the very similar Carboniferous formation that he had correctly identified in southern China, and it is erroneously represented as "Kohlenkalkstein" in his published maps of northern China. Lorenz* in 1902 and Blackwelder in 1903 independently recognized its Ordovician relation on the basis of few but characteristic fossils. Weller has described our collections.†

Sinian in general.—The development of the Sinian system in Shan-tung, where the Man-t'o red shales, the Kiu-lung group of green shale and varied limestone, and the Tsi-nan dolomitic limestone are distinguishable, is better known to us than the sequence of corresponding strata in any other province. The Man-t'o is present in northwest Chï-li and Shan-si, and is there followed by peculiar limestones like those of the Kiu-lung. Von Richthofen observed the same characteristic rocks in northern Ho-nan, latitude 34° 30'.‡ The red shale is wanting in the sections we saw on the Yang-tzï-kiang, but the conglomeratic and oolitic limestones are present. These varied relations led us to apply local names to the divisions which are unlike in different provinces. Their correlative significances are given in the following table:

System.	Shan-tung.	Liau-tung	Chï-li and Shan-si.	Ssï-ch'uan and Hu-peï.
Sinian	Tsi-nan limestone } Kiu-lung group } Man-t'o shale	Fu-chóu ? group Yung-ning sandstone	Ki-chóu limestone Man-t'o shale	Ki-sin-ling limestone Nan-t'ou tillite

The following paragraphs contain a summary of our observations regarding the Sinian in Shan-si and the Ki-sin-ling limestone of the Yang-tzï gorges, Hu-peï.§

In northern Shan-si, about the Wu-t'ai-shan, the Sinian is exposed in characteristic development, with the Man-t'o at the base and the cal-careous strata in great thickness above. The lowest stratum of the Man-t'o is frequently a conglomerate of pebbles of the subjacent rocks, and the materials of the formation are, throughout its thickness, coarser than is commonly the case in Shan-tung. The characteristic yellowish limestones occur, but generally high up and in two or three layers only. The total thickness is 180 to 335 feet, 55 to 100 meters.

Strata of gray to greenish shale and oolitic as well as conglomeratic limestone, aggregating about 600 feet, 180 meters, succeed the red rocks and correspond to part of the Kiu-lung group. Above these follow massive

---

* Beitrage zur Geologie und Paleontologie Shan-tungs, Lorenz, part 1.
† Vol. III of this work.
‡ China, vol II, p. 505.
§ For fuller details see vol. I, chapters VI and XII.

limestones, which are the equivalents of the Ch'au-mi-tién and Tsi-nan divisions and compare with them in massiveness. The total thickness of the system in northern Shan-si is not far from 4,000 feet, 1,200 meters.

On the middle Yang-tzï, in Hu-peï and eastern Ssï-ch'uan, the Sinian is represented by a limestone which is 4,500 to 5,000 feet, 1,350 to 1,500 meters, thick. The great formation is probably capable of subdivision on lithologic and paleontologic differences, but the distinctions are less marked than in Shan-tung or even in Shan-si. Regarding it as a whole it may be called the Sinian limestone, but to give it sharper definition we apply the name Ki-sin-ling, from the pass of that name where the provinces of Hu-peï, Ssï-ch'uan, and Shen-si corner. It is there typically exposed, and both Cambrian and Ordovician fossils were found in nearby sections.

At the base of the Ki-sin-ling limestone, at Nan-t'ou on the Yang-tzï, is a basal conglomerate of pebbles derived from the underlying tillite, embedded in greenish shale. It is but two feet thick and is succeeded by thin-bedded, shaly, oolitic, and in part chert-bearing limestones, 350 feet thick. We did not see the bottom of the formation in any other section, but the thin-bedded limestones probably appear in the gorges of the Yang-tzï and in the mountains near the Ki-sin-ling.

The upper and major part of the Sinian in the Yang-tzï sections is massive dark-gray or liver-colored limestone, free from chert. It is probably nearly 4,500 feet, 1,350 meters, thick. Farther north in the Ki-sin-ling pass, it consists of thinner bedded, more carbonaceous or bituminous strata. It is, however, in all its phases a great marine limestone.

The upper limit of the Ki-sin-ling is indeterminate, as there is a transition from it into interbedded shale and limestone, which passes upward into the overlying Sin-t'an shale. The transition strata are between 200 and 300 feet, 60 to 90 meters, thick, and at Sü-kia-pa in Ssï-ch'uan they yielded middle Ordovician (Trenton) fossils. The horizon is lithologically distinguished by a peculiar stratum of black chert or lydite, which was also found by von Richthofen and Lóczy in sections north of Kuan-yüan-hién, Ssï-ch'uan, 200 miles, 300 kilometers, northwest of Sü-kia-pa.

The age of the lower part of the Ki-sin-ling is not definitely fixed, as we found no fossils at Nan-t'ou, where the base is exposed; and those which we did find elsewhere were obtained from pebbles. They were picked up on a river bar of the Nan-kiang, near Chön-p'ing-hién, Shen-si, where strata of the Ki-sin-ling form the canyon walls, and they no doubt came from the immediate vicinity and from the lower part of the limestone. They comprise Lower Cambrian *Obolus asiatica* as well as lower Middle Cambrian types, and represent the top of the Man-t'o and base of the Kiu-lung group of Shan-tung.

The Ki-sin-ling thus includes at least upper Lower Cambrian at the base, and extends up to middle Ordovician (Trenton), at which horizon it passes by transition into shales, which are probably of Silurian or Devonian age.

Sinian strata are not definitely known by fossils to occur southeast of the Yang-tzï in southeastern China, but both von Richthofen and Lóczy,* with strong probability, refer certain strata, the Ta-ho grits and Lu-shan slates, to the period.

Sinian strata have not been recognized in characteristic limestones or identified by fossils in the Ts'in-ling-shan, yet they are in all probability present. Von Richthofen did not hold this view.† He wrote:

> The eastern Kuen-lung remained free from the Sinian transgression south of the northern base of the Fu-niu-shan and the Ts'in-ling-shan, and the adjacent region on the south.

Briefly the facts are as follows: In three sections across the Ts'in-ling range, observed by von Richthofen‡ and ourselves, metamorphic schists of the Wu-t'ai type are succeeded by a basal conglomerate, quartzitic rocks, and gray limestone. There is a great series of slates, and also carbonaceous limestone, which locally carries coal. Our observations of the relations of these strata are delineated on the geological atlas sheet *a*1, and while not conclusive, they indicate that the order of stratigraphic sequence is from the conglomerate, through quartzite, limestone, and slate, to the coal-bearing limestone. We agree with von Richthofen that the last-named is Carboniferous, but think it is distinct from the gray limestone, which corresponds with the Ki-sin-ling (Sinian) in position beneath a thickness of middle Paleozoic shales. If our understanding is correct, the Sinian is present in a mid-section of the Ts'in-ling-shan, near the northern base of the range.

There is but little knowledge regarding the Sinian system northwest of the Ts'in-ling-shan. Lóczy assigns various occurrences of limestone, which he describes as early Paleozoic, to the system, but as a rule, in the fact that they are highly siliceous, they much more nearly resemble the pre-Cambrian limestones of the Nan-k'ou system than they do the Sinian. However, in the vicinity of Qué-ta, latitude 36°, longitude 102°, west of Lan-chóu-fu, he observed in a small mountain range known as the "Cha-ji-shan" a heavily bedded limestone, in part dark and bituminous, in part lighter colored and somewhat interbedded with shale and sandstone, which exhibits an oolitic nodular structure such as is found also in the globulitic limestones of the Sinian. He regards this as a characteristic occurrence of the Sinian lime-

---

*Reise des Grafen Széchenyi, vol. I, p. 380, map.´
†China, vol. II, p. 713.
‡ *Ibid.*, vol. II, p. 565.

stone of sufficiently definite character to justify the identification of it and other limestones of the region as belonging to that system.

In the preceding pages the occurrence of the early Paleozoic (Sinian) strata has been described for various districts of China, from the province of Liau-tung, latitude 41°, to the Yang-tzï in latitude 30°; and from the eastern plains of the empire to the central Ts'in-ling-shan, in longitude 108°. It remains to consider the probably equivalent terranes which are known in the Himalayas.

In southwestern China rocks of early Paleozoic age, if they occur, are highly metamorphosed and have not yet been definitely recognized. It is not until we reach the central Himalayas, in longitude 80°, that we have any precise information. Thence westward, in the regions covered by Griesbach* and Hayden,† there exists, on the basis of fossils, definite evidence of the presence of the Upper Cambrian, which is underlain by a considerable thickness of conformable strata, presumably also of Cambrian age.

Griesbach calls these strata by the local name Haimantas, and gives the following classification for the eastern area:

In descending order:

Silurian:
    3. Series of quartz shales and slates.
    2. Shales and silky phyllites, with great thickness of quartzites.
Haimantas:
    1. Quartzite, generally purple, with great thickness of conglomerate.
Vaikritas and older gneiss. (Pre-Cambrian.)

The total thickness is estimated at about 4,000 feet, 1,200 meters. The base where the Haimantas come in contact with the Vaikritas exhibits no sharply defined plane of division:

At Milam, for instance, there is seemingly a very gradual passage from the micaceous schist south of that village, into greenish-gray phyllites and talcose schists with garnets of the Vaikritas, and finally into the thin-bedded quartzites, shales, and conglomerates of the Haimantas, and the change is so gradual that the boundary line could not be drawn with anything like accuracy.‡

Regarding the conglomerate, it is said (page 51):

The thick deposits of a coarse conglomerate and breccia are mostly made up of rolled and subangular fragments of rocks belonging to the crystalline area, and amongst them large boulders of quartzites and gneissose rocks seem to predominate. The matrix in which

---

* India Geological Survey Memoirs, vol. xxiii, Geology of the Central Himalayas, by C. L. Griesbach, 1891.
† Idem, vol. xxxvi, pt. i, Geology of Spiti, with parts of Bashahr and Rupshu, by H. H. Hayden, 1904.
‡ Idem, vol. xxiii, p. 51.

these boulders are firmly embedded is nearly always a hard flinty quartz rock, sometimes partially schistose. It is by far one of the most characteristic and easily recognized horizons of the central Himalayas, and is invariably met with in all Haimanta sections which I have seen.

Again on page 96:

This conglomerate, which in places strongly resembles a boulder-bed, merges into massive, intensely hard, dark-purple quartzites.

And on page 159:

The mineralogical character of the shales and quartzites lying below the typical purple quartzites with the boulder-bed (conglomerate) and the adjoining metamorphic schists, Vaikritas, merge into one another.

It is of interest to note the descriptive term, "boulder-beds," by which the conglomerates of the early Cambrian are designated, since we have discovered glacial deposits, probably of that age, on the Yang-tzï, in a region which, though 1,850 miles, 3,000 kilometers, distant, is practically in the same latitude, 31° north. However, the conglomerate was not regarded as a true boulder-bed of glacial origin by Griesbach, who, with reference to the ancient shore-line in the western Himalayas during the Haimanta age, states that he regards these deposits to be evidence of "a chain of elevations, from the waste of which the boulders and pebbles of the Haimanta conglomerate were derived." An inquiry addressed to Dr. Holland, the present director of the Indian Survey, has brought a negative, though perhaps not decisive, answer:

The question of the possible existence of Pre-Carboniferous glacial deposits in India is one which has recently received considerable attention. You are, no doubt, familiar with the published description of the Blaini boulder slate of the Simla area, the glacial origin of which is generally admitted. This formation was, until recently, regarded as possibly of Upper Paleozoic age, corresponding to the well-known Talchir and Salt Range boulder-beds; but there has been of late a general tendency to correlate the series of beds with which it is associated with the old, probably Pre-Cambrian, sediments of the Peninsula. I first drew attention to this in my General Report of the Geological Survey (1903–04) published in Records Geological Survey of India, vol. XXXII (page 156) and the conclusion based admittedly on negative evidence has received some support from the recent discoveries of Pre-Cambrian boulder slates both in Australia and in South Africa. In this connection, the occurrence of a typical boulder slate among the pre-Vindhyan rocks of the Son Valley in Rewa State is also of considerable interest (Memoirs, Geological Survey of India, vol. XXXI, p. 132); but it should be remembered that no striated boulders have been found in the Himalayas or in the Son Valley.

The possibility of the occurrence of a representative of the Blaini boulder slate in association with the fossiliferous beds of Gurhwal and Kumaon has been steadily kept in view during the progress of geological surveys in the Himalaya, and it may be assumed that, during his survey, Mr. Griesbach would have had this in mind when studying such a rock as the boulder-bed that he describes, and would have been on the lookout for evidences of glacial action. The same bed, however, is well known to Mr. C. S. Middlemiss, who had the opportunity of visiting many of the sections in Mr. Griesbach's company. Mr. Middlemiss states that the conglomerate bears no resemblance to a till, the matrix being a quartzite. The term "boulder-bed" has been applied to the Haimanta rock merely on account of the presence in it of unusually large boulders.

The middle and upper divisions of the Haimantas are thus characterized (pages 52 and 53):

The purple quartzites and conglomerates are in all sections overlain by a great thickness of bluish-gray phyllites, shales, and thicker bedded quartzites, traversed by many quartz veins. Towards the upper portion of it reddish-brown or pink quartz shales are intercalated. * * * The only fossil traces known from this system have been found in shales in this division. None of these organic remains are more than traces. They are: crinoid? stem impressions; bivalve? casts and numerous casts of Bellerophon sp. The latter occur both in the purplish-pink quartzite and in the shales accompanying it, and rather high up in the sequence of beds of this division.

In all the central Himalayan sections through the Haimantas, from the Kali river to the Spiti province, I have invariably found certain beds which constitute the third division. They consist of two zones of very hard quartz shales, the lower of which is formed by densely red and pink quartz shales, which pass upwards into greenish-gray quartzite and shales with pink shaly partings; the whole, as far as I know, quite unfossiliferous. Together these beds are not more than 250 to 500 feet in thickness.

In his account of the geology of Spiti (page 13) Hayden discusses differences of observations by himself and Griesbach, and bases his conclusion upon his own work, which in that region was the more thorough of the two. He fails to find any conglomerate which could be supposed to belong to Griesbach's lower Haimanta conglomerate, except in one instance in which the rock is clearly autoclastic. This difference may perhaps find its explanation in the suggestion of a glacial origin of the conglomerates, since they might in that case be of peculiarly local occurrence. After referring to the lower beds as slates, quartzites, and grits, Hayden proceeds (page 13):

The overlying beds, which presumably comprise Mr. Griesbach's upper Haimantas, consist of a series of black, purple, and gray slates, with gray, green, and red quartzites. The lower part of the series is chiefly argillaceous and the upper mainly siliceous. * * * Among the argillaceous beds are bands of an intensely black carbonaceous shale, resembling the carbonaceous shales of Simla. * * *

In the Parahio valley the upper siliceous beds pass up gradually into a series of gray and green micaceous quartzites and thinly foliated slates and shales, with narrow bands of light-gray dolomite.

The slates, which are usually dark blue or black, vary in composition from a soft argillaceous rock to a hard siliceous variety with much mica. * * * The slates are interbedded with great irregularity with gray, yellow, or whitish quartzites which are almost invariably capped by a narrow band of either calcareous quartzite or dolomitic limestone only a few inches in thickness. The limestone, which is gray on fresh fracture, weathers to a pinkish or brownish red and is again overlaid by slates which are at first argil-

laceous, but gradually become more and more siliceous till they pass up again into quartzites. This alternation continues with great regularity for many hundred feet. Towards
the top of the series the argillaceous beds give place to light-colored siliceous slates and
thin-bedded flaggy quartzites, with bands of red and pink dolomite, which latter gradually
increases in frequency and thickness till it becomes the predominant rock. These beds
constitute the oldest fossiliferous series hitherto found in Spiti.

After the statement of a detailed section, which comprises 1,188 feet,
371 meters, of strata, there follows the account of the fossils. Near the base
of the section, in a hard calcareous and micaceous quartzite, occur numerous
valves of a small brachiopod resembling *Lingulella*, with which are associated
indeterminate fragments of the head shields of a trilobite. About 400 feet,
120 meters, higher, trilobite remains are again abundant but poorly preserved. One hundred and sixty feet, 48 meters, still higher up, large numbers of specimens were collected in a very fair state of preservation. They
are said to consist chiefly of species of *Ptychoparia, Corda*, and allied genera.
A hundred feet, 30 meters, higher *Ptychoparia* is still found and with it large
numbers of fragments of *Olenus* sp. and *Dikelocephalus* sp. About 90 feet,
27 meters, still higher occur the uppermost determinable fossils, among
which *Ptychoparia* is rare and *Olenus* common. Although the collections
which are thus briefly described had not been studied, the forms recognized
are considered sufficient to warrant the inference that the fossiliferous beds
are of Upper, possibly also of Middle, Cambrian age.

POST-SINIAN DIASTROPHISM.

Sinian strata are commonly parallel in attitude with late Paleozoic
strata, where any such are present. The areas in which the sequence is
apparently immediate and sedimentation was continuous are limited, so
far as we know, to Central China; those strata that immediately succeed
are distinctly terrigenous sediments, which contrast decidedly with the
underlying marine limestones and represent a revival of erosion. Those
areas over which an unconformity of erosion without discordance exists
are extensive and comprise the Paleozoic of North China.

From these facts it follows that diastrophism at the close of the Sinian
time involved broad epeirogenic changes, without marked orogenic disturb-
anecs, within the provinces of eastern Asia.

An exact date of the movement can not be inferred in the region of
unconformity by erosion, for it is impossible to determine what strata may
have been removed from the latest which are there present beneath the
unconformity. The latest remaining are lower Ordovician; they are overlain by upper Carboniferous; and it is possible that Silurian, Devonian, or
lower Carboniferous deposits may have been spread over the area. It

is not, however, probable. Had such later strata accumulated to notable thickness over the several hundred thousand square miles under consideration, they must have been removed before the upper Carboniferous was laid down, and they should constitute a notable terrigenous formation between the Sinian and Wu-shan limestones in the region of continuous sedimentation. The Sin-t'an shale occupies this position, and it is but 1,800 feet, 550 meters, thick. Considering the extent of the lands from which the sand and clay were derived, the volume of the formation represents a comparatively thin sheet of material eroded. Further, so far as extensive, though not comprehensive, observations enable us to judge, the level of unconformity does not vary greatly from a general horizon in the upper Sinian; that is, the limestone remained practically flat and protected, was ultimately bared, and soon after was covered by upper Carboniferous. The changes of condition thus indicated are: (1) shallowing of the sea, such that the bottom, on which limestone had been accumulating, was scoured by marine currents; (2) after an indefinite time, gradual exposure of the limestone in a flat coastal plain, where at any one time a narrow belt was subject to erosion; and (3) progressive deposition of littoral or continental sediments upon the eroded surface. The last occurred not earlier than middle Carboniferous, and the other stages cover Devonian, Silurian, and upper Ordovician. It is possible to introduce a hypothetical stage of deposition in this series of events, between 1 and 2, if we assume that the deposit over the limestone was eroded in the gradual process of exposure; it is, indeed, probable that a thin formation of terrigenous character accumulated as a temporary storage of the material of the Sin-t'an formation in its passage from the crystalline rocks that yielded the quartz and ferruginous clay; but there is no direct evidence of any notable marine formation.

The date of inception of the epeirogenic movement may be fixed more nearly in the region of continuous sedimentation. Transition beds of shale and limestone follow the Sinian limestone, and among them occurs a stratum which contains a well-characterized middle Ordovician (Trenton) fauna. We may regard these transition beds as marking the close of the general marine condition and the beginning of the period of little or no deposition. In contrast to the areas of unconformity by erosion, the region of conformable deposits presumably remained somewhat deeply submerged, was not subject to marine scour by currents, and was so distant from the lowlands of the time that it received but little sediment. The evidence of this appears through consideration of the Middle Paleozoic strata.

If the preceding discussion proves trustworthy a very interesting parallel may be drawn with North America. The Cambro-Ordovician transgression there closes with an episode marked by withdrawal of the epicontinental sea from extensive areas within the continent, and in such wise that shallows and archipelagos of low flat islands took the place of wide mediterraneans. A similar physical change is indicated for eastern Asia. The physical history of both continental regions was one of prolonged submergence, favorable to cosmopolitan evolution, which resulted in the development of related faunas in the two areas; and the withdrawal of marine waters checked the faunal variation at the same stage on opposite sides of the globe. The geographic change was not occasioned by notable disturbances of the continental masses, although there probably was some gentle warping; but it may with reason be ascribed to a decided deepening of the ocean basins. The trend of the evidence, which may be traced more in detail in America and Europe than at present in Asia, is to establish proof of a general lowering of the sea-level, such a change as Suess designates a negative eustatic movement.* The world-wide prevalence of the preceding marine transgression, its long duration, the corresponding development of identical life conditions and consequently of closely related organisms, and the universality of the sea-level datum, all combine to render that particular negative movement one of the most definite time records in geologic history.

---

*La Face de la Terre, vol. II, p. 841.

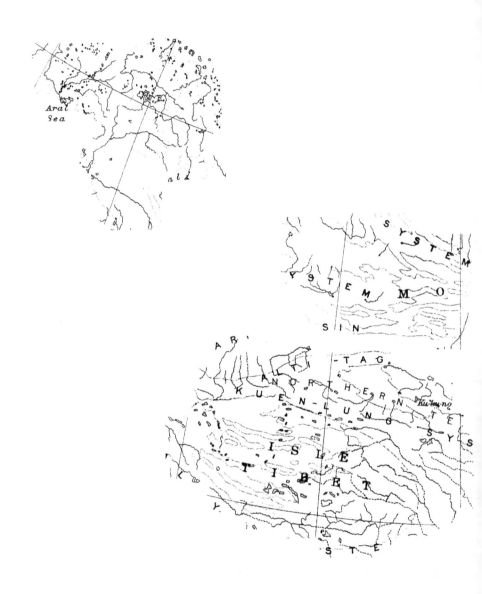

PLATE 4

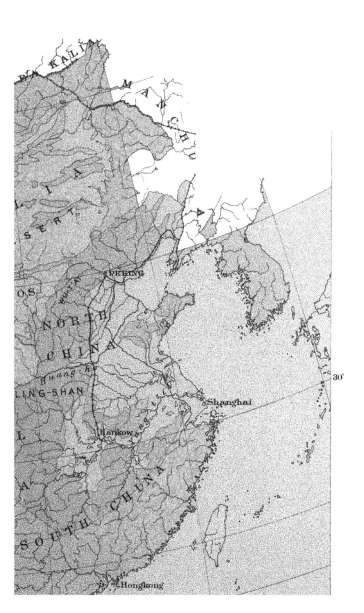

LEGEND

QUATERNARY
AND
TERTIARY
*(Basins deeply filled by
continental deposits)*

SINIAN STRATA
*(Areas within which they
are known to occur)*

SINIAN STRATA
*(Areas within which they
are supposed to occur)*

SINIAN LANDS
*(Areas where marine deposits
were probably not laid down)*

CHAPTER IV.—MIDDLE PALEOZOIC.

The preceding chapter upon the Sinian system deals with the stratigraphy of the Cambrian and lower Ordovician of eastern Asia. Under the division middle Paleozoic I propose to describe the strata which fall between the Sinian and the Carboniferous, comprising those assigned to the upper Ordovician, Silurian or Gothlandian, and Devonian.

We may first cite the early observations made by von Richthofen, who, in his second excursion into the geologically unknown China, explored the lower Yang-tzï, particularly between Kiu-kiang and Nan-king. The only account of his observations that I have yet been able to find is contained in the proceedings of the American Academy of Arts and Sciences, volume VIII, May 26, 1869. A number of formations are distinguished by local names. Of these the first four, namely, the Ta-ho, the Lu-shan schists, the Matsu limestone, and the granite intrusions, elude classification. The fifth in von Richthofen's enumeration, namely the Tung-ting sandstone, is also of uncertain position, but was supposed by him to extend conformably beneath the Devonian limestone, which is next in the series. To this Devonian representative von Richthofen gave the name of Si-ho, and he says of it:

> This is a limestone formation only 600 feet in thickness. The rock is full of chert nodules and contains numerous fossils, chiefly corals, encrinites, and brachiopods. *Aulopora repens* is of frequent occurrence among them, and other forms, too, indicate a Devonian age. The name is derived from a prominent hill generally known as Single Tree Hill, east of Nan-king, where I first found the fossils.

Seventh in the series named by von Richthofen are the Nan-king grits, which he described as a gritty and purely quartzose sandstone, mostly red but variegated, alternating frequently with a coarse conglomerate of perfectly rounded pebbles consisting exclusively of quartz. In certain dark shales which are interstratified occur indeterminable fossil plants.

Conformably overlying the preceding is the Carboniferous limestone, called Ki-tau, a name which takes precedence of all others for this formation in China. It consists of three members in this locality, namely (1) a lower limestone characterized by *Fusulina cylindrica*, 1,400 feet, 420 meters; (2) black sandy shales, black lydite, and soft sandstone, about 400 feet, 120 meters, thick, containing in the lower strata especially large specimens of

53

Productus semireticulatus, accompanied by numerous other brachiopods, bivalves, corals, and fenestellas; and (3) an upper limestone more than 1,600 feet, 480 meters, thick, which is separated from a bed of coal of the preceding member only by a thin stratum of black shale and is similar in nature to the lower limestone. The Ki-tau corresponds in a general way with the Wu-shan limestone.

The further description deals with beds of Mesozoic and later age which rest unconformably upon the Carboniferous.

A type of the middle Paleozoic is found in central China, in eastern Ssï-ch'uan, southern Shän-si, and western Hu-peï. The region is mountainous, the strata are folded, and exposures are practically continuous in the superb canyons of the Yang-tzï-kiang and its tributaries.

Descriptions of supposed Devonian rocks in place are given by Pumpelly in his account of the Yang-tzï gorges. The great limestone which rests on the pre-Cambrian rocks, and is now known to be of Cambro-Ordovician age, was called by him Devonian on the supposition that the fossils of that age from China were from the limestone in question. In fact, Devonian in that region is either absent or represented by the shale which appears immediately above the limestone in several places in the gorges through which Pumpelly passed in his adventurous journey.*

Sections observed by our expedition occur on the Ta-ning-ho, longitude 110° east, and on the Yang-tzï-kiang, from that meridian east to I-chang.† At the base collections of middle Ordovician fossils were obtained from the transition strata of the Ki-sin-ling limestone near Sü-kia-pa, a village on the Ta-ning-ho in eastern Ssï-ch'uan, latitude 31° 40', longitude 109° 40'. From a local bed at the top, a pink limestone near Ta-miau-ssï, we obtained forms which may be Devonian or lower Carboniferous.

At Sü-kia-pa the middle Paleozoic strata lie in a closely folded syncline, with nearly vertical dip between outcrops of the underlying Ki-sin-ling, (Sinian) limestone. The sequence is continuous, conformable, and apparently complex; the following are the principal rocks in order from the top downward:

> Massive gray limestone, with abundant black chert nodules and upper Carboniferous fossils; base of the Wu-shan limestone.
> Sandy shale, chiefly green, in part black; calcareous, ferruginous, bituminous, and coaly near the top; also bituminous and black 200 to 300 feet, 60 to 90 meters, above the base; weathers earthy brown and iron stained: total 1,800 feet, 550 meters; Silurian, Devonian, or lower Carboniferous fossils at the top.
> Gray limestone, 200 feet, 60 meters.

---

* Smithsonian Contributions to Knowledge, vol. xv, Geological Researches in China, Mongolia, and Japan, during the years 1862–65, R. Pumpelly, pages 5–6.
† Vol. 1, pp. 270–271; atlas sheets *d* 6 and *d* 7, geology.

Black chert, "lydite"; a characteristic stratum which does not, however, always occur in its place; checks on weathering into small angular fragments; thickness 5 feet, 1.5 meters.

Green shale with numerous nodules of gray to brown ferruginous limestone: 100 feet, 30 meters; highly fossiliferous, middle Ordovician (Trenton) fauna.

Light-gray to bluish limestone, in layers 2 to 6 inches thick, containing *Orthoceras* and coiled gastropods, lower Ordovician; upper part of the Sinian (Ki-sin-ling) limestone.

The characteristic green shales of the middle Paleozoic occur in typical development on the Yang-tzï-kiang, in longitude 110° 45′ E., at the village of Sin-t'an, and the name is conveniently used to designate the sequence of strata between the Ki-sin-ling (Sinian) limestone below and the Wu-shan (Carboniferous) above. Transition beds 200 feet, 60 meters, thick connect the Ki-sin-ling and Sin-t'an and are characterized by upper Ordovician fossils. The top of the Sin-t'an is marked by a distinct but conformable contact of shale with the overlying massive Wu-shan limestone.

From the limits stated it will be seen that the Sin-t'an formation represents all of Silurian (Gothlandian), Devonian, and lower Carboniferous time. Including the upper Ordovician transition beds, it is but 2,000 feet, 600 meters, thick, and as a whole is a rather monotonous clayey, sandy deposit, modified by more or less calcareous, ferruginous, or bituminous admixture. We were not able to detect any evidence of unconformity at the base or top of the formation, nor within it. We regard it as a unit in a complete sequence of strata; but it was evident that sedimentation was meager during most of the periods represented. While it is possible that the deposit accumulated chiefly during some particular epoch, we have no clue to that one.

The fauna found in the limestone nodules in green shale, and in light-gray limestone near by is described by Weller in volume III of this series. It affords a sound basis for correlation with the Trenton limestone of the United States and the Vaginoceras limestone of the Baltic, the relations being closer with the latter. The horizon falls in the middle Ordovician.

Although we repeatedly crossed the strata equivalent to this fossiliferous bed, both north and south of Sü-kia-pa, none of the party observed the fossils at any other locality. We may have overlooked them, especially in running the swift current of the Ta-ning-ho and Yang-tzï-kiang in boats, but it is probable that the fossiliferous calcareous nodules in the green shale are of occasional, not general, occurrence. Where they do occur the fossils are conspicuous.

In general the Sin-t'an formation is apparently unfossiliferous. Its bituminous phases have much the character of the middle Devonian shales of the eastern United States, and where green and sandy it suggests the Chemung.

At the top, in a pink limestone, Blackwelder found a few fossils which Girty and Bassler have studied.* They are not able to agree on a definite correlation, but state that, whereas the bryozoa indicate an age corresponding to the American Mississippian, the brachiopods do not resemble any Mississippian fauna, and if they stood alone would lead Girty to class them as possibly Silurian (Gothlandian). The presumption is that the horizon is at least late Devonian or lower Carboniferous, but if so the fauna is unlike others of those times previously collected in China.

This stratum of pink limestone lies but a few feet below the Wu-shan limestone and 1,500 feet, 450 meters, below a thin bed of anthracite, next to which Blackwelder collected a Chonetes of upper Carboniferous affinities.

Silurian and Devonian strata in central China were first described by von Richthofen, who came upon them in southwestern Shen-si and northern Ssï-ch'uan, en route from Han-chung-fu, longitude 107° 12', latitude 32° 50' to Kuan-yüan-hién, longitude 106° 15', latitude 32° 20'. They there form the southernmost ranges of the Ts'in-ling-shan, north of the Red Basin of Ssï-ch'uan, and are intricately folded and overthrust. The strata are, however, richly fossiliferous, and have consequently been well identified.

South of the town of Ning-kiang-chóu von Richthofen crossed an anticlinal range, overturned toward the south, from which he obtained the following section.† (The strata are designated in the accompanying figures and discussion by the letters *l* to *e* in ascending order.)

(*l*) Limestone alternating with other strata, namely, gray, splintery, siliceous limestone, 15 feet; clay shales; lydite conglomerate, and lydite sandstone. The pebbles of the conglomerate in some of the beds are much rounded, in others they are rounded on the corners, and in still others they are quite angular and sharp-edged. The sandstones are practically a fine-grained conglomerate of the same kind. Thereupon follow, in continuous alternation, shales and limestones in groups of varying thicknesses. Characteristic are: red limestones, red calcareous marls which break up in friable shaly pieces, and also green, somewhat crystalline limestone, yellow and green calcareous marls, etc.

The limestone layers are full of well-preserved fossils; in the upper parts trilobites and brachiopods predominate (among them *Orthis calligramma* and *Spirifer radiatus*); in the lower, crinoid stems and corals. Dr. Kayser determines the age as middle Silurian approximately equivalent to the upper Llandovery.

Adjacent to the exposures of *l* and underlying that terrane in the overturned limb of the anticline comes a formation *k* and others in sequence, as follows:

(*k*) A great thickness of green clay shales, with occasional thin limestone layers, in which brachiopods occur.

---

*Vol. I, pp. 273–274.
†China, vol. II, p. 596.

(*i*) Beds of limestone which contain much clay shale; the limestone consists almost entirely of corals in distinct, very well-preserved individuals, which attain more than one foot in diameter and are separated by shaly material. *Favosites forbesi, Halysites catenularius*, several kinds of *Heliolites, Cyathophyllum, Amplexus*, and *Alveolites* occur, according to the determination of Lindström, together with other Silurian forms. Crinoids are numerous, brachiopods and *Orthoceras* scarce. Of trilobites I did not find any.

(*h*) Green clay shales, which are distinguished by occasional layers full of nodules of limestone in bedded form. When weathered these are recognized as the remains of corals, which belong to some of the species enumerated in *i*, but are much smaller. Limestone layers of the most varied character and coloring are interbedded with the shales and alternate with them, now in thick beds, again in the thinnest layers. The number of fossils is extraordinarily great in this easily recognized formation. Every calcareous piece exhibits organic structure. Together with the corals occur innumerable long crinoid stems, small trilobites, and many brachiopods, especially species of *Orthis*. The last occur in such heaps that they alone constitute single limestone beds. According to the determination of Dr. Kayser, the terranes *i* and *h* are Upper Silurian, approximately equivalent to the Wenlock.

(*g*) Gray limestone and soft gray calcareous marl, which, according to the occurrence of *Atrypa reticularis*, is Devonian.

(*f*) Strongly bituminous limestone of blackish and liver-brown color, which is without question identical with the Carboniferous limestone.

The gray, splintery limestone at the base of this section is not recognized by von Richthofen as representing the Sinian, but probably is of the transition beds, which follow the typical limestone of that system on the Ta-ning-ho. As von Richthofen had not distinguished the Sinian from the Carboniferous, he was not prepared to recognize it where only the top appears in an unfamiliar stratum.

The lydite conglomerate corresponds with the black chert bed which occupies a similar horizon in the Ta-ning-ho section, and the fauna collected by von Richthofen from the beds *l* and *k*, and described by Kayser, is regarded by Weller as closely related to that which we obtained at Sü-kia-pa.

In this connection it is desirable also to mention certain fossils described by Martelli,* collected by the missionary Giuseppe Giraldi in the vicinity of "Lean-san," in the Ts'in-ling mountains of southern Shan-si.† I have not been able to find the locality "Lean-san" (which may perhaps be transliterated Liang-shan, Two Mountains) on any European or Chinese map available, but it can not be far from the section observed by von Richthofen. Weller regards the fauna collected by Father Giraldi as even more nearly related to that from Sü-kia-pa than is that described by Kayser.

---

* Bol. Soc. Geol. Ital , vol. XX, p. 295, Fossili del Siluriano Inferiore dello Schensi (Cina), Alessandro Martelli.

† Probably the same as Abbé David's Lean-chan, described by Paul Fischer in Bull. de la Soc. Géol., Ser. III, vol. II, p. 408.

Von Richthofen's section includes Silurian (Gothlandian) strata as determined by abundant fossils (beds *i* and *h*) and a very meager representation of the Devonian (bed *g*). The expedition of 1903–04 found no certainly Silurian fossils in the Ta-ning-ho section. Lóczy, who traversed the Kia-ling-kiang just west of von Richthofen's locality, passed in a boat and did not observe the Silurian; but at a point somewhat further north he obtained a definite observation of the occurrence of the Devonian. The locality is in the western Ts'in-ling-shan, in the province of Shen-si, at the village of Paj-suj-kiang (Pai-shui-kiang, White Water River). Lóczy described the occurrence as follows:*

When we had left a steep terrace [Talstufe] our way entered a narrow ravine, in which steep limestone walls rose on both sides almost to Paj-suj-kiang. In this gorge the limonitic sandstone, and the clay shale which is associated with it, is repeatedly exposed wherever the arches of the anticlines have been cut through by the action of the river. Above the sandstone occur well-bedded, thin, limestone layers, and above these the thick beds of the bituminous limestone extend to the high ragged edge of the ravine.

In the marly strata between the dark-blue limestone layers occur numerous petrified remains of shellfish, such as corals, crinoids, brachiopods, fenestellas, and also great heaps of shells of a large form of mussel. The cross-sections of the latter remind one strikingly of those of megalodonts. In the material which was here collected the following species were determinable:

Spirifer, cf. canalifera Valen.　　　　Cypricardinia lamellosa Phill.
Dielasma (cf. sacculus) minimum n. sp.　　Favosites cervicornis Goldf.
Pentamerus brevirostris Phill.　　　　Stromatopora sp. indet.
Megalodon (sp.) v. Pachydomus? sp.

In Europe these forms are characteristic of the Devonian system. The thin Devonian strata pass upward gradually into a thick bituminous limestone; in consequence of this relation, because of its extraordinary thickness, and also on account of its analogy with the strata which von Richthofen described and definitely determined to be Carboniferous, I place this limestone formation in that system. We may therefore conclude, from the stratigraphic relations as well as from the conditions of deposition, that, on the southern slope of the Shï-ta-shan, the Devonian and Carboniferous occur in an unbroken, continuous sequence.

Our route extended about 14 kilometers upon the strata which I have just described; below occurred the carbonaceous sandstone and clay shale; above that bluish limestone interbedded with black chert, in the marly layers of which occurred the Devonian fossils, and finally uppermost came the Carboniferous limestone, which, with a thickness of about 400 meters, extended to the tower-like pinnacled divides and ragged cavernous plateau of the Shï-ta-shan.

In the paragraphs preceding the above-quoted description, Lóczy gives an account of the rocks immediately below those containing the Devonian fossils, describing them as gray ferruginous sandstones, containing pyrite and brown iron oxide with interbedded threads of coal. This

---

*Reise des Grafen Széchenyi, vol. 1, p. 433.

description agrees with that of certain portions of the Sin-t'an formation observed by our expedition in eastern Ssï-ch'uan, and the position of the fossiliferous limestone corresponds with that of the pink limestone in which, near Ta-miau-ssï, Blackwelder found the few fossils that are doubtfully referred by Girty to the Devonian or lower Carboniferous.

On the agreement of the observations of von Richthofen and Lóczy and ourselves, we may conclude that throughout northern Ssï-ch'uan and southern Shen-si the Devonian is represented by thin strata of calcareous, marly, bituminous character, which nowhere attain very great thickness, which may in fact be wanting in some sections, and which are conformable to the Silurian (Gothlandian) below and the Carboniferous above.

In connection with the occurrence of middle Paleozoic fossiliferous strata in the western Ts'in-ling-shan it is desirable to consider the formations of the Heï-shui series, which were observed by our expedition in crossing that range in longitude 108° and which are tentatively assigned to the period. They are siliceous argillites of greenish and reddish tints, which more nearly resemble the Sin-t'an formation than any other that we know, and appear to occur as it does, between a massive limestone below and a coal-bearing limestone above. In general terms the stratigraphic similarity between this sequence and that of the Paleozoic in the middle Yang-tzï region is such that we have little doubt in classing the whole as also Paleozoic and the slate formation as middle Paleozoic. The slates include quartzite and conglomerate at the base and appear to be thicker than their supposed equivalents further south. We regard them, therefore, as a nearer shore phase, and also recognize that they may include a more complete sequence than is found in the Sin-t'an.

MIDDLE PALEOZOIC STRATA BEYOND CHINA.

We may next consider the distribution of Devonian strata in northern and western Tibet, as determined by Obrutchov, Mushketov, Romanovski, Bogdanovitch, and Stoliczka.

The journey of Lóczy, otherwise so fruitful in geologic details, gives no clue to the occurrence of the Devonian in the ranges of northern Tibet. In justice to him it must be said that his opportunities for observation were limited by the official attitude of the Chinese authorities and the conditions attending the expedition to which he was attached. Although he traveled during four months within a short distance of the northern base of the Nan-shan, he was able to observe the rocks in place only in two localities; in the vicinity of Kan-chóu-fu and south of Sü-chóu. He describes strata which he considers to be early Paleozoic, but which, as stated on a previous page, are not identified by fossils and probably are of pre-Cambrian age. His observations are described in the twelfth chapter, volume I, of his work, on pages 552 to 559, but need not here be further considered. Obrutchov's journeys in northern Tibet were very extended. His observations are unfortunately not fully available. The two great volumes, Central Asia and Northern China, published in Russian, are in effect an orderly transcription of his field-notes, from which it is difficult to obtain an idea of geologic relations.* In his preface Obrutchov reserves all conclusions for a third volume, which is not yet available. He has, however, found a most able interpreter in Suess, who supplemented the data found in the volumes by correspondence with the author. I quote from him. The region referred to lies in the heart of Asia, south of the Gobi desert and north of the Tibetan plateau, between latitudes 35° and 40° north, and on either side of longitude 100° east. The description includes an account of rocks which are both older and younger than middle Paleozoic and in this volume are systematically discussed elsewhere, but the quotation is best given as a whole.

Suess says:†

The width of the Nan-shan, between the northern margin of the Tsai-dam and the southern margin of the oasis of Kan-su, is equal to the width of the Alps between Biella and Freiburg, and one may say that all the region commonly designated under the name of Nan-shan corresponds approximately in extent to the Swiss Alps. The great height of the ranges is accompanied, however, by a relatively even more striking altitude of the valleys, a circumstance which diminishes the differences of level in the interior of the mountainous region, but the observer is even thus below the limit of eternal snow.

From Prevalsky to Roborovsky numerous but hardy observers have contributed to knowledge of the topography of this mountainous region, but up to the present time only two competent geologists have traversed it. The first was Lóczy, whose researches outside of the northern border were strictly limited to the route from Lan-chóu to Liang-chóu, to the valley of Si-ning and the southeastern part of the country. He was followed by Obrutchov, who studied several sections across the northern ranges and in a long expedition traversed all the chains in the vicinity of their northwestern extremity. Thence, following their direction, he reached the field of observation of Lóczy, south of the Ku-ku-nor, and thereafter, crossing the eastern part of the ranges, regained the oasis of Kan-su.

After commenting on the reconnaissance character of these investigations and the incompleteness of our knowledge, he says:‡

In the meridian of Sü-chóu four high mountain chains rise between the oasis of Kan-su and the longitudinal valley of Ku-ku-nor. They are uniformly directed west-northwest. They are: the Richthofen range, the To-lai-shan, the range of Alexander the Third, and a fourth chain which is little known and is separated from the preceding by a

* In examining these volumes I have been greatly assisted by Mr. G. Zon, assistant forester, Bureau of Forestry, U. S. Department of Agriculture.
† La Face de la Terre, French edition, 1902; vol. III, part I, page 231.
‡ Ibid., page 232.

Valley of the Sou-lai-khé [Su-leï-ho].* In the continuation of the Alexander mountains to the west-northwest, or a little to the south of this line, rises the Ta-sioné-shan [Ta-sué-shan, Great Snowy Range]. The Richthofen range is 50 to 60 Versts across and may be divided into several groups of folds, of which the first without doubt exceeds 20,000 feet, 6,000 meters. At the northern base, near Tsin-fo-sy, southeast of Sou-tchéou [Sü-chóu], there is a mass of granite, but although the granite rises rapidly it does not extend far into the mountains. It is followed by a band of upper Carboniferous with beds of coal, inclined to the southwest; then by lower Carboniferous with *Productus striatus;* by a red and green DeVonian sandstone; and finally by quartzites and limestones which are probably Silurian. Before the first summits of the range are attained the beds are already inclined towards the north, and all the ridges of the Richthofen range may be considered as closely appressed folds of analogous Paleozoic terranes, which are tilted to verticality or overturned toward the south. There seem also to be overthrusts. Possibly some of the coal-measures belong to the Angara series. At the southernmost divide called the Tsin-pin-ta-pan (14,000 feet, 4,220 meters) and even above this pass strata of the Gobi series lie in discordance against the southern base of the Richthofen mountains; they everywhere dip very steeply toward the Valley of Khoun-tschoui, which forms the limit of the slope of the To-lai-shan. * * *

Still further to the west-northwest on the middle course of the Sou-lai-khé [Su-leï-ho] there occur in the Paleozoic series of the Richthofen mountains red and green sandstones, in the midst of which are interbedded layers which contain DeVonian fossils (*Rhynchonella alinensis* according to Tschernyschew). Obrutchov mentions in another locality *Spirifer elegans* and *Spirifer anosoffi.* [Central Asia, II, pp. 9 and 10.]

We may here interpolate an extract from Obrutchov's own account :†

On the southern slope [of the Richthofen range] we see a thick series of supra-Carboniferous deposits forming one of the southern ridges, the divide of the range. They strike north-northwest diagonally to the range, and dip steeply on both sides. Below the peak Yang-k'ou-er appear more ancient formations: gray sandstone and shales slightly metamorphosed, which I consider ancient Paleozoic. They do not resemble closely the ordinary metamorphic sandstone and schists of the Nan-shan and other parts of central Asia. They may be Silurian, or perhaps still older. Fossils have not been found. This series is much contorted in various directions, east-northeast, east-west, and west-northwest. Apparently it formed the range when the supra-Carboniferous sandstones were deposited at its southern base; now it forms the southern slope of the main ridge. The remaining part of the Richthofen range represents an enormous anticline, which is overturned southward, and on the limbs is composed of a great thickness, 2,800 to 3,500 feet, 850 to 1,050 meters, of conglomerates, mostly of red color; in the southern limb they stand almost perpendicular and in the northern they dip inward. Near the axis of the fold the conglomerates are replaced by red and green sandstones and shales with thin layers of conglomerate. These strata apparently form secondary steep folds. The strike is west-northwest, parallel to the modern Richthofen range, as might be expected, since the range has been formed in its present position by the dislocation of this series. The thickness of the conglomerate of the southern limb forms the crest of the mountain chain, with its

---

*The French spelling khé is a transcription of the Russian хе, pronounced as хо, and rendered in English ho, the Chinese for river.

†Central Asia, North China, and the Nan-shan, vol. II, p. 145.

steep rocky peaks. The series of red conglomerates underlain with red and green sand-stone and shales is most probably pre-Carboniferous. I think it equivalent to the series of red and green sandstone with middle Devonian fossils, which I have found in the Richt-hofen range considerably further west on the section from the spring A-tza-k'ou to the oasis Ch'óu-ma-er (pages 8–10); and also in a nearer section, in the canyon of Tsin-fo-ssi, east of the meridian of Sü-chóu. We shall meet a series of red and green sandstones and shales, which underlie Carboniferous limestones with fossils and which form the northern ridges of the range; a similar red conglomerate was seen in the eastern part of the range Mo-ma-shan. Hence it is more probable that all these similar formations of the Richt-hofen range present shore and shallow water deposits of the same age, namely Devonian.

Suess continues:*

The To-laï-shan in the meridian of Sou-tchéou is a somewhat narrow chain, but like the Richthofen range, rises high above the snow-line, of which the altitude is 14,600 to 15,200 feet, 4,400 to 4,600 meters, in this region. It is distinguished from the preceding range by an outcrop of gneiss along its southern border. The attitude of the beds, accord-ing to the description of Obrutchov, is very peculiar: after having crossed, in coming from the north, the intermontane belt of the deposits of Gobi, one observes intense folding in a Paleozoic series, which is very thick. Locally the strike is northeast or north-northeast; then the strata resume the dominant west-northwest orientation, and finally the most recent formations, the Carboniferous with beds of coal and Fusulina limestone, rest upon the gneiss. The impression which the section gives in this locality is indeed not so much that there is an overlap of the upper Carboniferous upon the gneiss as that there is an overthrust toward the south, of the entire Paleozoic series upon the gneissic belt.

Referring to another section of the To-laï-shan, about 50 versts, 53 kilometers, west-northwest of the first† Suess describes the continuation of the folds which involve Devonian and later Paleozoic strata resting upon metamorphic schists; and shows further that the gneissoid zone along the southern margin, which has widened considerably, appears to be *underlain* both north and south by upper Carboniferous coal-measures. This struc-ture presents a problem of overlap or overthrust which remains indeter-minate.

The range of Alexander the Third rises south of the broad valley of To-laï-kouan, which is entirely filled with debris; it is formed, as may be seen from two transverse sections taken a score of versts apart, of lower Paleozoic strata upturned and folded. Toward the south these beds pass beneath a great syncline of Carboniferous strata, in the midst of which the sandstones above the coal-measures are very extensively developed. This synclinorium is so broad that the sandstones overlying the coal-measures constitute a series of secondary arches, and the summits of the Ou-jé-chan, the most elevated part of the chain, probably consist of them.‡

Obrutchov himself sums up his description of the section of the range of Alexander the Third by saying that "To-day's observations show that

---

*La Face de la Terre, vol. III, p. 234.
† *Ibid.*, bottom of page 235.
‡ *Ibid.*, page 236.

the series of gray schists, quartzites, and limestones, is considered older than that of gray, green, and lilac shales, which we take for Devonian." These older schists, quartzites, and limestones belong to that series which has been considered early Paleozoic, but which I correlate tentatively with the late Proterozoic. I shall refer to it again in connection with Paleozoic diastrophism.

Having referred briefly to the structure of the fourth chain, which is but imperfectly known, but which is undoubtedly formed by beds of Paleozoic strata, in part at least, Suess proceeds:*

The structure of these four ranges of the Nan-shan, from the oasis of Kan-su to the vicinity of the Valley of Bouk-haïn-gol thus presents the following characteristics: gneiss is seen only upon the southern flank of the To-laï-shan; all the rest of the mountainous region is formed of closely folded Paleozoic sediments. Near the northern margin there are suggestions of an overturn toward the north; in the interior of the mountains the beds appear in a vertical position or exhibit a movement toward the south. The gneissic zone is not associated with the lowest strata of the Paleozoic series, as would be expected in the normal succession, but instead with the Upper Carboniferous.

The observations of Obrutchov in the southeastern portion of the Nan-shan, as quoted by Suess, show that the pre-Carboniferous Paleozoic strata are present in that part of the range. There can be no question but that outcrops of Devonian strata extend from the vicinity of Lan-chóu southeastward along the trend of the mountain chains to southwestern Shen-si, in the region where Lóczy collected fossils of that period near Paj-suj-kiang.

Proceeding westward from the Nan-shan range, along the southern margin of the Gobi, we have in the western Kuen-lung a section by Bogdanovitch across the ranges adjacent to what is known as the Valley of the Winds. I again quote Suess:†

One of the sections which Bogdanovitch observed across the western Kuen-lung crosses the Valley of the Winds. Coming from the north the first chain, the Altyn-tagh, is formed of Devonian with masses of granite, exactly as in the Russian range which lies far to the southwest. The second chain, the Youssoup-alyk-tagh, which follows the Tchi-men-tagh, is a broad band of gneiss. The Valley of the Winds corresponds to a Carboniferous syncline. This section, which somewhat further west is extended southward, first comes to the opposite limb of the syncline, then to the steeply inclined Devonian and the great granite massif of the Kyzyl-oungouïn-in-tiouré, beyond which there succeed further outcrops of Devonian strata. It is in the region beyond this massif that there rises the Aīlik-tagh, where Bogdanovitch found at a great altitude "polypiers" of middle Devonian age.‡

---

*La Face de la Terre, vol. III, page 237.
† Ibid., page 243.
‡ See also Beitrage zur Stratigraphie Zentral Asiens, Suess in Denkschriften der k. Acad. der Wiss., Wien, LXI, 1894, p. 435.

Devonian strata are known where the Kuen-lung converges in the Mustagata, toward the Tién-shan, and have been identified in the southern ranges of that great system, north of the Taklamakan desert. Farther to the northeast, in Trans-Baikalia, they constitute the oldest fossiliferous rocks known.

We have thus traced the repeated outcrops of middle Paleozoic strata from the province of Shen-si in central China, in longitude 106° east of Greenwich, northwest and west to longitude 74°. The terranes probably comprise pre-Devonian Paleozoics, at least as far west as longitude 94°, but in the section last quoted from Bogdanovitch middle Devonian appears to rest directly upon the crystallines. Nevertheless, in the present state of geologic knowledge of these remote regions, it would be hazardous to assert that earlier Paleozoic, even to Ordovician or Cambrian, is not represented. Somewhat farther west, in the Zaravchan (Sarawschan) range, longitude 70°, Romanovski collected *Halysites catenulatus.* The middle Paleozoic of eastern Turkestan is closely related to and connected with that of Europe by way of the Urals, and also with North America by Siberia.*

The line of observations which we have followed from Central China pursues the course of one of the great mountain trends of Asia, between the Gobi and the Tibetan masses. Paleozoic strata may probably be absent over part of the latter, at least if the inference based on structural axes be valid. Around its eastern end the ranges extend southward toward the Malay peninsula, and in Burmah they are joined by the Himalaya trends.† We may trace the occurrence of Middle Paleozoic rocks along this southern route also.

Devonian is known from several localities in southwestern Ssï-ch'uan and adjacent districts in Yün-nan, among the Paleozoic strata which form the mountain ranges of western Tibet, bordering the Mesozoic red basin on the west. Westward from Ja-chóu-fu (latitude 30°, longitude 103°), in the vicinity of "Lin-tschin-shien," Lóczy found the Jurassic itself thrown into folds, resting unconformably on strata which he assigns in general terms to Silurian, *i. e.*, early Paleozoic. He refers to the rocks simply as limestone, and makes the correlation without paleontologic evidence.‡ Beneath these strata occur granite and diabase. A few miles further west, between the Fu-yung-ho and the village of Hoani-pu (Hua-ling-pu) he crossed the Ta-shian-ling, which is composed of granite and quartz porphyry and flanked on both sides by coal-bearing Jurassic strata. On the northeastern slope the supposed Silurian beds occur. The town of

---

* De Lapparent: Traité de Géologie, 1906, vol. ii, p. 864.
† Suess: La Face de la Terre, vol. iii, map.
‡ Reise des Grafen Széchenyi, vol. i, p. 677.

Ni-tou, a short distance to the northwest, lies in a synclinal basin of early Paleozoic strata overlain by Rhætic and Jurassic and flanked by ranges of granite and quartz porphyry. In this vicinity, at the village of Hua-ling-pu, Lóczy observed closely folded shales and dark, bituminous, earthy limestones, from which he collected a number of fossils of middle Devonian age.* The fossiliferous limestone beds were thin and apparently underlain by sandy shales and gray, half-crystalline limestones, and the presence of Silurian strata beneath the Devonian was observed.

While the fossils collected in this locality leave no question of the presence of Devonian strata, the relations which appear to be implied in the accompanying diagrams, between the Paleozoics and the adjacent granites, are not at all clear. One would infer that the granites were pre-Paleozoic; on the other hand the descriptions suggest that they may be intrusives and of post-Paleozoic age.

In northern Yün-nan, in latitude 28°, longitude 104°, is the celebrated locality of Ta-kuan-ting, from which von Richthofen obtained the rich collection described by Kayser.† Von Richthofen did not himself collect the fossils, but purchased them in the province of Ssï-ch'uan, whither they are brought in great numbers as medicinal charms.

La Touche and Datta determined the presence of Devonian strata in the northern Shan states, where they found a full and characteristic fauna. The presence of the Silurian rocks is probable, but not determined.‡

The most recent and complete account of the Ordovician and Silurian and Devonian in the Himalayas is that given by Hayden.§ He describes 650 feet, 195 meters, of strata, comprising shaly sandstones at the base (150 feet, 45 meters), shaly, dolomitic, and siliceous limestones (420 feet, 125 meters) among which six divisions are distinguished, and reddish quartzite (80 feet, 24 meters); the last is followed by the unfossiliferous white Muth quartzite. The strata have yielded fossils ranging from Caradoc to Llandovery or Wenlock, and therefore corresponding in general position with the beds of middle Ordovician and Silurian age in central China. The white Muth quartzite passes upward into siliceous limestone, which is considered to be Devonian, and Devonian strata are probably present throughout the Himalayas, although of very moderate thickness. The western extension of this series along the ranges of western Asia to the Caucasus is not clearly made out, but is most probable.‖

* Reise des Grafen Széchenyi, vol. I, p. 682.
† China, vol. IV, p. 76.
‡ Noetling: Gen. Report G. S. I., 1900.
§ Memoirs G. S. I., vol. XXXVI, part I, pp. 24–27, 1904.
‖ De Lapparent: Traité de Géologie, 1906, vol. II, p. 812.

These notes suffice to trace the marine connection of Central China with Europe by a route south of the Tibetan plateau region, and thus to bring out the fact that the interior sea of Asia, the Tethys, during middle Paleozoic time completely surrounded the plateau region, which was probably a land area.

MIDDLE PALEOZOIC DIASTROPHISM.

Diastrophic movements in China during the middle Paleozoic were very slight. As has been brought out in discussing the sedimentation of the Silurian and Devonian periods, there was no considerable deposition of marine sediment, no evident accumulation of continental deposits, no notable depth of erosion. A stable condition of the southeastern continental masses is plainly indicated.

A similar inference holds for Gondwana Land in the peninsula of India and the Tibetan mass, so far as we may draw one from the meager sedimentation that represents Silurian (Gothlandian) and Devonian in the Himalayas: a few hundred feet of limestone and quartzite; the lands were not high.

A somewhat different suggestion lies in the Devonian and Silurian (?) deposits of northern Tibet, which Obrutchov, the explorer of the Nan-shan system, describes as quartzites and shales of great thickness. They appear to represent an epoch of vigorous denudation, and their volume seems to stand for an elevation equivalent to a mountain range. According to Bogdanovitch, there is an overlap of middle and upper Devonian onto an eroded surface which exposes granite; the elevation had, therefore, by middle Devonian time, given place to peneplanation and subsidence; he regards this invasion of the sea as an event of prime importance and designates it the Kuen-lung transgression.

It is not clear what the nature of the elevation was: an orogenic movement, perhaps accompanied by granitic intrusion, or an upwarp without folding or visible intrusion? Comparison with the Taconic disturbance in New England or the Caledonian movement in Scotland, both during Silurian time, tempts speculation to postulate a similar orogenic event in central Asia; but the conservative position taken by Suess* in deferring a correlation of these events is the sound one in the limited state of knowledge. This attitude is the more reasonable because the Taconic and Caledonian disturbances are geographically and presumably causally related to the Atlantic basin, whereas the region in central Asia belongs to a distinct continental province.

---

* Beitrage zur Stratigraphie Zentral Asiens, in Denkschriften der k. Akad. Wiss., LXI, 1894, p. 435.

PRE-CARBONIFEROUS UNCONFORMITY.

Throughout northeastern China, in the provinces of Chï-li, Shan-si, Shan-tung, and Ho-nan, there is an unconformity which brings the Ordovician in contact with the Carboniferous. Wherever it has been seen the underlying terrane consists of the upper Sinian (lower Ordovician) limestone; whereas the overlying strata are shales which lie a hundred feet or so below beds that carry upper Carboniferous fossils, either marine shells or plant remains. The hiatus appears,·therefore, to represent later Ordovician, Silurian, Devonian, and lower Carboniferous times.

Local observations all agree that the Ordovician and Carboniferous strata are strictly conformable; so closely parallel are they that von Richthofen, who repeatedly saw the contact, assumed that they formed an unbroken sequence, and characterized the Sinian limestone as Kohlenkalk. He did not find any of the sparsely distributed fossils, by which its age has since been determined by Lorenz and ourselves, and his error is one which sprang naturally from the fact that he had previously observed a massive Carboniferous limestone in South China. It follows from the conformity of dip over several hundred square miles that the region in question was not subjected to disturbing influences during the periods represented. The Sinian strata were not folded; neither, during all this time, were they raised to an elevation at which they might have been deeply eroded.

Apart from the fact that the strata above and below the plane of unconformity are identified by fossils and the existence of a hiatus is thus proved, the contact itself shows evidences of erosion. The Sinian limestone is deeply fissured and cavernous. The cavities show waterworn or corroded surfaces, commonly found in limestones exposed to meteoric waters. These cavities are filled with bright-colored clays, which owe their varied hues to the oxidation and hydration of the iron which they contain. Bog-iron ore, sufficient in quantity to be mined, and· calcareous tufa are found at the contacts.

The significance of these facts has been discussed in connection with the consideration of post-Sinian diastrophism. For detailed descriptions of the unconformity as we saw it, reference may be made to volume I, pages 48 and 147. In describing the coal-field of Po-shan in northern Shan-tung, von Richthofen* states with reference to the Kohlenkalk (Sinian limestone) that he found no fossils, but noted that in contrast to similar limestones in other regions it contained no chert. This distinction is one in which the Sinian differs from the Carboniferous limestone of southern China.

---

*China, vol. II, p. 203.

He then proceeds:

There follow, in a relation which I am not able to determine clearly, but apparently conformably, a series of deposits which begin with bright-colored clays. They are in part stratified, in part without recognizable structure, and fill inequalities in the eroded dolomite.

Farther north, in the province of Liau-tung, von Richthofen describes the sequence in the coal-field of T'ai-p'ing,* and gives the following section:

Limestone, which is heavily bedded and characterized by yellow markings that resemble corals but are indistinct in outline on the kidney-brown background, and which does not contain flint, is followed by yellow dolomitic limestone, thin-bedded yellow dolomite with limonite, red ferruginous clays with kidneys and nodules of hematite and limonite, white sandstone, and coal.

On the margin of the great coal-field of southeastern Shan-si the explorer ascended from Hwai-king-fu to the coal-measure plateau and observed the contact of the coal-bearing T'ai-yang series with an underlying limestone in the type locality. He says:†

Coal and iron district of T'ai-yang. Throughout a long distance on the west side of the Pi-lo-shan the loess hides the strata above the limestone; but before one reaches T'ai-yang one has opportunity to see them in beautiful exposures, for there the firm rock forms a flat ridge which connects the limestone range with the mountains that rise further west. They consist of an alternation of sandstone, bright clays, yellow dolomite, rauchwacke, and limestone with flint nodules. I was particularly struck with the analogy of the system in this region with that which occurs at Po-shan-hién in Shan-tung, between the limestone and the coal-bearing sandstone. Here, as well as there, the clays are used for making pottery. In both localities the iron ores are related to the dolomites. They do not, however, occur interbedded with them, as indeed the rocks of this horizon are in general not clearly stratified, but exhibit the most striking irregularities in short distances. Here the surface of the dolomite is full of caverns without order, and in these the iron ores are especially common.

The figures on page 410 of the volume illustrate the observed relations between the coal-measures and the underlying limestone, and show the two to be conformable in dip except in one locality, figure 86, which is, however, not described in the text.

The observations which von Richthofen may have made in the course of his journey from Canton to Han-kow have unfortunately not been published. He crossed the extensive coal-fields of Hunan and particularly examined that on the Lui-ho, a branch of the great Hsiang-kiang, which flows into Tung-ting lake. From the fact that coal-measure plants occur in abundance‡ we infer that the Carboniferous strata resemble those of the

*China, vol. II, p. 287.
†Ibid., vol. II, p. 411.
‡Ibid., vol. I, p. XXXVI.

northern provinces in being of continental rather than marine character. As the Carboniferous limestone occurs in the region to the northwest and west, and also to the east between Kiu-kiang and Nan-king, Hu-nan appears to represent a peninsula extending southwestward into the southern sea. It is probable that the unconformity which defines the northern land areas may have extended thus far south.

The land area which is indicated by this marked unconformity was probably a southeastern extension of the continental area which existed in Mongolia and central Asia. We turn naturally to the observations of Lóczy in his journey from Si-an-fu to Lan-chóu-fu, to ascertain the conditions at the base of the Carboniferous in the province of Kan-su, but unfortunately his route over the loess plateau avoided the exposures which are to be found only in the deeper canyons, and he did not see the base of the Carboniferous. The section of the Lo-pan-shan* is the only one which may have bearing upon the unconformity in question. The exact locality is west of Pa-liang-fu, near longitude 106°, latitude 35° 30'. He describes a limestone, which he calls Kohlenkalk, but which may be upper Sinian, as being unconformably overlain by bright-colored sandstones, marbles, and conglomerates, which he assigns to the Ober-Karbon, i. e., Permian. If the underlying limestone is indeed the Carboniferous, the base of that system is not seen. If, on the other hand, it be the Sinian, the Carboniferous is wanting and is here overlapped by the Permian.

---

*Reise des Grafen Széchenyi, vol. i, p. 486.

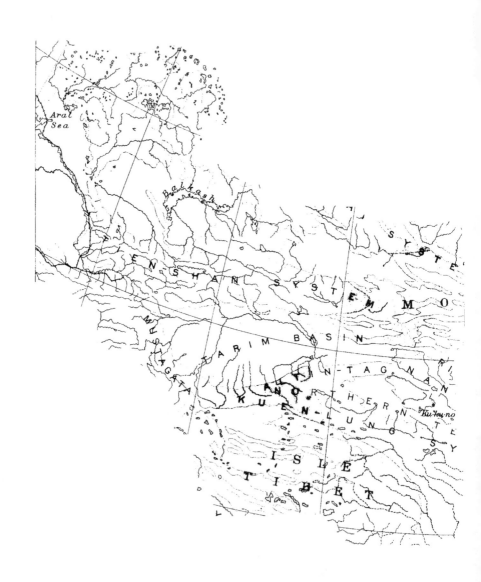

PLATE 5

LEGEND

QUATERNARY
AND
TERTIARY
(Basins deeply filled by
continental deposits)

MIDDLE PALEOZOIC STRATA,
CHIEFLY DEVONIAN
(Areas within which they
are known to occur)

MIDDLE PALEOZOIC STRATA,
CHIEFLY DEVONIAN
(Areas within which they
are supposed to occur)

MIDDLE PALEOZOIC LANDS
(Areas where marine deposits
were probably not laid down)

CHAPTER V.—LATE PALEOZOIC.

CARBONIFEROUS STRATA.

The Carboniferous strata of China constitute two very distinct series, both of which are, however, apparently of upper Carboniferous (Pennsylvanian) age. The one is a sequence of essentially continental deposits: sandstone and shales, with numerous coal-beds and occasional thin bituminous limestones. Many of these strata are no doubt marine, the limestones in particular being characterized by marine fossils; but many others belong to a coastal plain and are either seashore or fluviatile or marsh deposits. In China they extend over the northern provinces, Liau-tung, Chï-li, Shan-tung, and Shan-si, and there is probably a large peninsula in the province of Hu-nan. Beyond China they are of general occurrence in the ranges of Mongolia and Tibet, which rise through the great mantle of desert deposits.

The other series is in fact a great single formation, a limestone which in some places is more than 4,000 feet, 1,200 meters, thick and is characterized by marine fossils throughout. It has been identified south of the Ts'in-ling-shan in the provinces of Shen-si, Ssï-ch'uan, and Hu-peï, and further east along the Yang-tzï-kiang in An-hui and Kiang-si. It occurs in the Kuén-lung and Nan-shan mountains of Tibet, and also in the Tién-shan in Turkestan. It is furthermore a wide-spread formation throughout all of southwestern China, Burmah, and the Malay peninsula. The marine deposits are also extensively represented in the Himalayas.

I first take up the account of the marine formations, which conformably succeed the Devonian, and next the interbedded continental and marine formations which rest unconformably upon older rocks.

MARINE CARBONIFEROUS.

The first description of the Carboniferous limestone of southern China was furnished by von Richthofen, who gave it the name of the Ki-tau limestone, from a locality on the Yang-tzï-kiang. His description has been given in connection with his account of the Devonian of the same region.

Three hundred and seventy miles, 600 kilometers, west of von Richthofen's locality, in the gorges of the Yang-tzï-kiang, between I-chang and

Wu-shan-hién, the Carboniferous limestone is beautifully exposed. Its distribution is shown on the geological map of the Yang-tzï gorges, plate XXXVI of volume I of this report. It was also repeatedly crossed in the sections on the Ta-ning-ho in eastern Ssï-ch'uan, and its relation to the underlying 'strata was thus repeatedly observed by the members of the expedition of 1903-04. The greenish, sandy, and bituminous shales of the middle Paleozoic are followed by calcareous shales, and these by massive gray limestones in which layers and nodules of black flint constitute a conspicuous feature. A pink limestone at the top of middle Paleozoic strata yielded a few fossils already referred to and described in volumes I and III of this report. They are not earlier than Silurian (Gothlandian), may probably be Devonian, and may even be as late as the lower Carboniferous (Mississippian). This fauna has nothing in common with that from the base of the gray limestones which closely overlie it, and it is evident that the life conditions underwent a notable change between the two epochs. Nevertheless there is such a uniformity of sequence in repeated sections that the strata can not be otherwise described than as conformable. We are thus led to conclude that in this district the sequence of deposits which began with the lower Cambrian (Sinian) limestone, and continued conformably through the middle Paleozoic, remained unbroken through the lower, middle, and upper Carboniferous. Yet it is to be noted that the sedimentation from late Ordovician to upper Carboniferous time was very meager and may possibly have ceased altogether at intervals in the middle Paleozoic.

This upper Carboniferous limestone, which we distinguished by the name Wu-shan from its prominent section in the Wu-shan gorge of the Yang-tzï, is probably the same as the Ki-tau limestone of von Richthofen. But until the faunas have been more fully collected and the geology of the entire region more accurately mapped, it is as well that they should be designated by separate names.

The Wu-shan limestone exhibits a sequence which has been described by Blackwelder in volume I, p. 275.

In the district in central China which has just been described we have, so far as the faunal evidence goes, a large development of the upper Carboniferous and only a very meager representation, if any, of lower Carboniferous strata. But from this locality toward the northwest and southwest Carboniferous limestones may be traced through the observations of von Richthofen, Lóczy, Obrutchov, and others, and they are apparently not restricted to the upper Carboniferous. De Lapparent gives a good synopsis* of the occurrence of the older Carboniferous strata. In view of the

---

* Traité de Géologie, fifth edition, 1906, p. 914.

very meager collections which have been obtained it is, however, hardly worth while to attempt to distinguish the different horizons of the great limestone formation in the light of present knowledge. And therefore, in describing its distribution I shall refer simply to the Carboniferous, which may include both lower and upper or either one of them.

Northwestward from the Yang-tzï gorges Carboniferous limestone is known to form conspicuous ranges in northeastern Ssï-ch'uan and southern Shen-si,* and through the work of Obrutchov† and Lóczy‡ it has been traced into northeastern Tibet, where it is more particularly represented in the southern mountain chains of the Nan-shan. In the section observed by Lóczy near Paj-suj-kiang the transition from fossiliferous Devonian strata to the gray Carboniferous limestone is accurately described as one of strict conformity, and this is also true of the sections given by von Richthofen in the same district of northern Ssï-ch'uan. The formation in this district might readily be confused with a similar limestone of Permian age, which rests unconformably on the folded Paleozoics.

The expedition of 1903–04 found the Carboniferous limestone and argil_lite constituting a recognizable terrane in the valley of the Han above Hing-an-fu and in the Ts'in-ling-shan, in longitude 108° east. The characteristic rocks are highly carbonaceous; they vary from limestone and quartzite to black slate. Their Carboniferous age is not determined by fossils, as they are more or less metamorphosed, but it is inferred with confidence on the basis of their commonly carbonaceous character and their stratigraphic relations.§

The relations of the Carboniferous limestone to older rocks in northern Tibet are not clearly expressed in the available literature. Such a section as that of Obrutchov's of the Richthofen range‖ is indefinite. The Carboniferous is represented as lying between folded masses of rocks that are questionably assigned to the Silurian, it resting unconformably upon the one mass and being overthrust by another. The structure itself is unusual, as it is delineated, and the age of the Silurian rocks being doubtful it is not possible to say what the unconformity means. If the older strata are those of the Nan-shan sandstone and belong to the latest pre-Cambrian (Hu-t'o) system, as has been suggested in the chapter on that subject, the Carboniferous in this northern range overlaps upon pre-Paleozoic

---

*China, vol. II, p. 599.

†Central Asia and the Nan-shan Mountains, Obrutchov, vol. II, p. 356, pl. II (in Russian). Also, La Face de la Terre, E. Suess, vol. III, p. 271.

‡Reise des Grafen Széchenyi, vol. I, p. 433.

§See chapter on the Han province, vol. I, p. 300.

‖Central Asia and the Nan-shan Mountains, vol II, p. 164, fig. 130. Also La Face de la Terre, Suess, vol. III, p. 232.

(*cb*) Dark coal-bearing shale, with slickens covered with calcite. The thickness of this bed could not be estimated closely on account of loose stuff on the slope, but I could see by the position of the tunnels that below this shale occurred

(*cc*) the principal coal-bed. From the height of the slope the thickness of *cb* and *cc* may be estimated at 18 to 20 meters.

(*d*) Dark or black bituminous clay marls, with fine calcite veins and full of fossils; including also yellowish-gray layers of mussel shells. I saw these only on the dumps; the thickness is therefore unknown, and I can only infer that they occur below the coal-bed.

(*e*) Gray, bituminous, marly limestone, which occurs on the southern slope from the mines and in the foot of the principal mine; which alternates with thin layers of coal, and which is full of fossils.

(*f*) Yellow sandstone, colored with iron hydroxide and in part shaly.

The thickness of the strata included from *d* to *f* is uncertain, as the base of *f* was not visible; but all the strata seen on the southern side of the basin lie unconformably on the Nan-shan sandstone, which dips steeply toward the south.

Lóczy gives a list of fossils which serve to correlate the strata with the Russian Upper Carboniferous (Mjatschkowo).

There are many references in the works of Obrutchov to coal-bearing strata in Mongolia and the Nan-shan. In some cases they are identified as Carboniferous, and in others as supra-Carboniferous. The conditions which prevailed in Shan-si were general throughout central Asia north of the Tibetan arm of the Tethys, and the deposition of continental deposits containing coal, yet interbedded with occasional marine limestones, was wide-spread. The area thus characterized reached into Turkestan on the west, perhaps, as far as longitude 80° east, but was bounded by a sea on the northwest, where the Carboniferous limestone of the Tién-shan mountain system was being deposited.

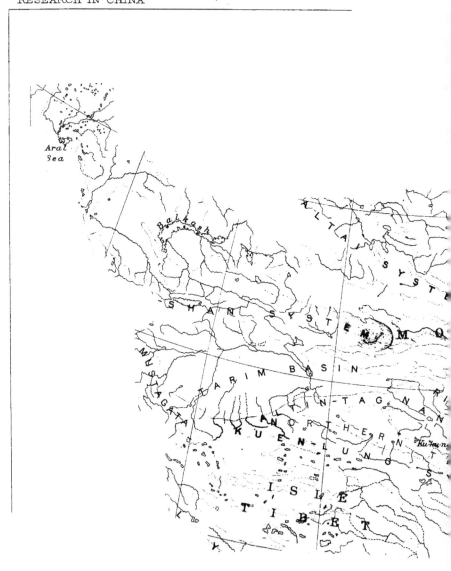

PLATE 6

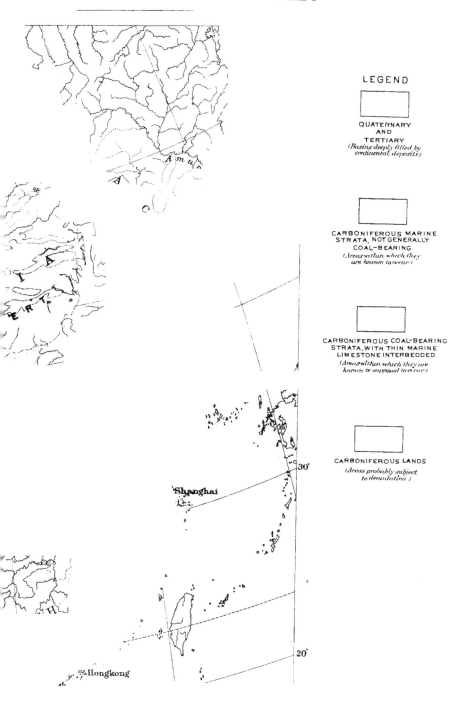

LEGEND

QUATERNARY
AND
TERTIARY
(Basins deeply filled by
continental deposits)

CARBONIFEROUS MARINE
STRATA, NOT GENERALLY
COAL-BEARING
(Areas within which they
are known to occur)

CARBONIFEROUS COAL-BEARING
STRATA, WITH THIN MARINE
LIMESTONE INTERBEDDED
(Areas within which they are
known or supposed to occur)

CARBONIFEROUS LANDS
(Areas probably subject
to denudation)

CHAPTER VI.—PERMO-MESOZOIC.

Theoretical considerations.—There is no sharp plane of division to be discovered between the Paleozoic and Mesozoic in Asia. In some districts there was transition of sediment and of faunas; elsewhere, during equivalent time, there is disturbance of physical or biological conditions. In one place a decided change of sediment, in another an equally abrupt change of fauna, in a third an unconformity of dip between beds that differ in kind and in fossils: these may be taken as local dividing planes; but they do not range themselves into one horizon. They fall anywhere within the later Carboniferous, or within the Permian, or Trias. It is in the nature of things that a great transformation of features, climates, and faunas, such as characterized the passage from the period of the Old Life to that of the Middle Life, should take time, and that the evidences of change should appear at different geologic moments in different places. The better to understand the facts which are presently to be listed, we may review briefly the theoretical sequence of events.

The leading fact that distinguishes the early Mesozoic from the late Paleozoic is the contrast in the extent of the lands, which, through withdrawal of the epicontinental seas, became relatively very wide. This change theoretically resulted from enlargement of the ocean basins by subsidence and widening, which was associated with protuberance of the continents and local deformation of rocks by folding.*

There is no question about the leading fact, of which Asia, as well as North America, offers a striking illustration. The wide seas of the late Carboniferous shrank, in the Permian and Triassic, to embayments around continental platforms. One effect of that shrinking was interference of emerging lands with circulation of marine currents. Hence followed climatic differentiation and contrasts of warm and cold currents in the seas and sharper contrasts of climatic zones on lands. A further effect of these changes was the development of new species and extinction of organisms incident to altered conditions of habitat. Faunal change was evidently a late effect, since it was sequential upon the others. Do these phenomena afford any precise datum that might serve to distinguish a definite epoch,

---

*Geology, Chamberlin and Salisbury, vol. II, p. 656.

as the last of the Paleozoic or the first of the Mesozoic, an epoch that could be fixed upon as contemporaneous throughout several continents, or even within one extensive continent? It does not appear that they do.

Granting that deepening of sea basins would withdraw the continental seas simultaneously if the continents stood fast, it is evident that the lands would emerge soon in one area and only much later in another if the contineuts were here rising and there sinking, as was the case. The effects of emergence, *i. e.*, erosion or continental aggradation, were therefore not even approximately contemporaneous, for we must bear in mind that earth movements are slow. Denudation and aggradation were among the earliest processes to become active and the longest to continue. Among the Asiatic effects are: that eroded surface which uncovers strata ranging from the Carboniferous down to the Silurian in the western Himalayas;* or those extensive continental deposits of red cross-bedded sandstones and associated shales, which in North China conformably overlie the Upper Carboniferous coal-measures.

The differentiation of local climates from the preexisting conditions of a general, somewhat uniform climate may reasonably be compared with similar changes during the Pliocene and Pleistocene, following upon the general mildness of the Tertiary. The Paleozoic-Mesozoic transition presents even more remarkable extremes, such as the wide occurrence and persistence of the Gondwana flora and the development of centers of glaciation in India and Australia.† Regarded as an effect of refrigeration and aridity, the formation of iron oxide in quantity can not have been an immediate result; geographic and meteorologic changes had progressed notably before the deposition of red rocks could have become general.

Marine organisms, evolving in ocean waters (that in contrast to air constitute a medium which is extremely conservative in maintaining the conditions of life) are long protected against change and also very sensitive to change, especially of temperature. Thus it is not surprising that in some ocean currents where the mildness of Carboniferous seas was maintained, Paleozoic forms should have lived on, even after their habitat had been invaded by Mesozoic types, as witness the Productus limestone of India; or that elsewhere, on the diversion of warm waters and occupation of their territory by chillier ones, there should be speedy extinction, even of a whole fauna. Even though the endurance of the adult be considerable, that of the larva is very slight, and the latter is the critical factor. When we thus consider the physical conditions which permit or limit the existence of species, the value of fossils as evidence of contemporaneity is

---

*Hayden: Memoirs Geological Survey of India, xxxvi, pt. i, p. 52.
† Textbook of Geology, A. Geikie, vol. ii, pp. 1058, 1079, and numerous references there given.

qualified by the geographic phase during which the organisms lived. In the world-wide Carboniferous seas marine organisms attained cosmopolitan distribution; and correlation between Europe, Asia, and America rests upon a firm paleontologic basis; but when those seas, so far as their epi-continental parts were concerned, became transformed into bays, straits, carribbeans, and mediterraneans, organisms met with great variety of experience. Where favored they persisted or slowly evolved; where adversely conditioned they died out or varied, and cosmopolitan relationships gave place to provincial diversity. That they did so sooner here and later there, and that the old sometimes returned to a habitat from which it had been driven during a temporary unfavorable condition, is the record of stratigraphy and paleontology.* It follows that correlations on a paleontologic basis are less reliable during the transition from Paleozoic to Mesozoic than during those periods of more general conditions which preceded and followed; and that even though Permian and Triassic faunas were more abundant and better known than they are in Asia, we should still have difficulty in establishing a plane with reference to which, in central Asia, China, Australia, and India, we could say: that which is beneath is Paleozoic and that which is above is Mesozoic. There is no such general plane.

Unconformity is a condition common to a transition stage, regarding which a word is necessary in explanation of the diversities which Chinese sections present. The term covers a great range of phenomena indicating discontinuity of deposition, from that which may result from nondeposition, marine scour, or subaerial erosion, without obvious disturbance, to the most striking differences of structure. We are apt to think of a submerged surface as one subject to sedimentation, or *vice versa* of a surface which has not received sediment as one which must have been above water; but the inference is not valid in view of the fact that marine currents, when confined in straits or shallows to such an extent that the water next the bottom moves, are capable of carrying sediment past a district, or even of scouring the bottom. Considering the work of corrasion done by rivers whose depth and bottom pressure are relatively insignificant, the capacity of marine currents to scour can not be questioned, provided the waters are equipped with abrading materials in the form of silt, fine sand, or siliceous spicules.† Nondeposition during an indefinite time or corrasion of an undisturbed stratum may result. It is thus that we may explain the contrast of strata where the Carboniferous limestone is concordantly overlain by red terrigenous deposits, as in eastern Ssï-ch'uan. The limestone

---

* Williams: Devonian Section of Ithaca, New York, Journal of Geology, vol. xv, p. III, October, 1906.
† Agassiz: Three Cruises of the Blake, vol. I, pp. 136–139.

which, after the manner of lime deposits, presumably hardened in course of accumulating, may be supposed to have been swept clean in a shallowing gulf or strait until the current was checked and the terrigenous sediments from nearby land were laid down. It is evident that the strata in contact with the limestone may be nearly continuous with it or indefinitely younger than it.

On the other hand, apparently at the other extreme of unconformity, we have strata extended across eroded folds, with marked discordance of dip. ; In a case described by von Richthofen and Lóczy near Kuan-yüan-hién in northeastern Ssï-ch'uan, erosion has laid bare Silurian strata on an anticline including several thousand feet of Paleozoics, and the overlapping formation is possibly Permian, more probably Triassic limestone. There is no doubt of folding, which resulted in a more or less elevated range, accompanied by erosion which reduced that range to a peneplain. The obscure feature is the answer to the question: how does the unconformity consist with the fact that in nearly adjacent regions strata are concordant in dip, from Carboniferous to Jurassic; such being the case on the middle Yangtzï, in eastern Ssï-ch'uan, 200 miles, 320 kilometers, distant.* I take it that the phenomena illustrate two things which are by no means peculiar to Central China. The first is that effects of folding are localized in linear belts and may be very decided in certain zones which lie adjacent to others that do not share in the disturbance; which remain, for instance, the bottoms of deep synclinoria. And the other is that the growth and wasting of an elevation due to folding takes no longer than the transition period from Paleozoic to Mesozoic. How long that was it is difficult to say, but in geologic terms it included the latest Carboniferous, the Permian, and part of the Trias, in Asia; and probably exceeded Pliocene and Pleistocene time, judging by the effects of mountain growth and planation.

PERMO-TRIASSIC STRATA.

Angara series.—There are in Asia two series of continental deposits which are approximately contemporaneous, but are geographically distinct. They both include members which range in age from Permian to Lower Jurassic, and thus cover the Paleozoic-Mesozoic transition period. The one is the well-known Gondwana series of the Indian peninsula, the other the Angara series of Siberia and the Altai region.† Representatives of the Gondwana series have been identified in Indo-China and also in Australia. These districts lie south of the great central mediterranean of Asia, to which Suess has given the name of *Tethys*, whereas the occurrences of the Angara

---

* Research in China, vol. I, chapter XIII, pp. 285 et seq.
† La Face de la Terre, Suess, vol. III, p. 27.

series lie north of that sea, on the continent of Angara. It is probable that the continental area extended from the Gobi region of central Asia, eastward to the present coast line or beyond it, and southward beyond the site of the Ts'in-ling-shan, south of which it was bounded by the Tethys. We thus place in the Angara series the terranes which are about to be described.

Strata which are assigned provisionally to horizons ranging from Permian to Rhætic, inclusive, are widely distributed in China, from Shan-tung on the east to Mongolia on the west and from Shan-si in the north to Indo-China in the south. The identification of the coal-bearing series as Rhætic rests on fossil plants and is usually qualified by an alternative assignment to Lower Jurassic. Without attempting to prejudge the definite correlation these formations are discussed in this section.

In central Shan-tung observations by the expedition of 1903–04 with reference to Permo-Triassic strata were made in the vicinity of Lai-wu-hién, and are described in volume I, in the chapter on stratigraphy under the Sin-t'an District. They are much less complete than is desirable, since our attention was given chiefly to the older rocks, and in the sections which we could conveniently observe there were gaps covered by alluvium at those points where we should expect the passage from known Carboniferous to supposed Permian. Carboniferous beds, identified by marine fossils in a bituminous limestone interbedded in the coal-bearing sandstones and shales, were followed by red cross-bedded sandstones. On account of their color we have regarded the latter as Permian, and we take them to be of fluviatile origin. Succeeding these, but in interrupted succession, occur red, black, and greenish shales interbedded with basaltic flows. There is abundant evidence that throughout an area including practically all the central part of Shan-tung there was a volcanic district from which eruptions were numerous and extensive. The activity began probably during the latest Paleozoic and continued well into the Mesozoic. In volume I these volcanics have been classed with the Carboniferous, that systematic term being extended to cover the Permian.

Overlying the volcanic series near Yen-chuang, Shan-tung, occur beds designated by us as the Sin-t'ai series, and by von Richthofen and Lorenz assigned to the Jurassic. In the Sin-t'ai area they are exposed with a thickness of several thousand feet, as gray to red sandstones and shales with thin beds of conglomerate. Von Richthofen included with them the adjoining coal-bearing rocks of Ts'ai-kia-chuang ("Tsing-ko-tschwang"), but on the evidence of the plants collected Schenck placed the coal-measures of that locality in the Carboniferous.

In the coal basin of Weï-hién von Richthofen observed the occurrence of coal-beds in close proximity to granite, an exceptional relation, since the deposits he had seen previously were in strata conformable to his Kohlenkalk (Sinian). Those were, however, of Carboniferous age, whereas these may probably be of Lower Jurassic, according to plants collected by Lorenz and described by Potonie.

Throughout northwestern China, from the eastern outcrops of the Carboniferous coal-measures west of Peking and the western margin of the great plains, the area of red and yellow sandstones overlying the coal-measures is very extensive. Von Richthofen called these strata "Ueberkohlensandsteine" or "Plateauschichten," on account of their relation to the coal-measures, their indeterminate age, and their occurrence in the extensive synclinoria which form the elevated districts of Shan-si. He gives but one section,* in which he distinguishes from below upward the following members:

(1) Sandstone and shale of the coal-measures with several small coal-beds.

(2) The principal anthracite bed which is accompanied by black slates.

(3) A series of mostly dark shales and soft yellow sandstones.

·(4) Reddish yellow clayey shales with occasional thin sandstone layers.

(5) Green and red shales, which are very thin-bedded and hard and break up into fine debris. These are interbedded with some thin sandstones which form benches.

(6) Soft and also hard sandstones, heavily bedded; 100 feet thick.

(7) Bright-colored shales like those in 5, which also alternate with a few sandstone beds. The principal colors are green, gray, blue, and red; thickness, 120 feet.

(8) Firm sandstones of a greenish color, very evenly bedded with white mica; in part containing quartz grains as large as peas, which also occur of still larger size and form solid conglomerates.

(9) Shales and sandstones of dominantly red color.

The last member is probably not the top of the series, the total thickness of which is estimated at 3,000 feet, whereas that of the underlying coal-measures (which are included in the preceding section) is said to vary from 400 to 600 feet.†

At Ta-tung-fu, northern Shan-si,‡ is a sequence of strata which is peculiar not only in being remote from other similar basins, but in the great thickness of strata and of the individual coal-beds. Von Richthofen describes it as follows:

Lower Jura; coarse quartzose sandstones with clayey matrix, interbedded with white strata of tripoli and also of siliceous calcareous formation, both containing remains of plant stems; also coal-beds, which are associated with black shales. The visible thickness is 1,200 feet. The character and color of the strata vary continually. The strike is north 30° east, the dip is in a direction north 60° west, at first at an angle of 40°, which lessens to 10°, and the upper beds lie almost flat.

---

*China, vol. II, pp. 414–415.

† Ibid., page 453.

‡ Ibid., vol. II, pp. 356–359.

The strata below the Jurassic are Sinian, and the relation is no doubt one of unconformity by erosion, but the stratification is approximately parallel in the Sinian and Jurassic beds, and the contact was not seen in the section described in the above quotation. In an adjoining section, however, near "Heï-ku-tsze,"* the unconformable contact of the two series was observed and is represented in the diagrammatic section as one of unconformity of dip as well as of erosion.

CENTRAL CHINA.

Under this head I consider the occurrence of Fermo-Mesozoic strata along the Yang-tzï, in the Ts'in-ling-shan, and in the Red Basin of Ssï-ch'uan.

I begin the enumeration of the known occurrences with that which is at once the most eastern and also the first recorded, the deposits in the valleys adjacent to the Yang-tzï, between Han-kow and Nan-king, as described by von Richthofen, who says:†

Sanghu sandstone and conglomerate.—The deposition of the Kitau limestone ended with a considerable disturbance, as the next formation follows quite unconformably. It consists of a quartzose sandstone and quartzose conglomerate, interstratified with thick layers of red clay, and carries a coal-bed at a place sixty miles below Hankau. Black shales, which overlie the coal, carry some remains of plants. I was unable to establish the thickness of this formation.

Commencement of the outbreaks of porphyry.—The porphyritic eruptions have probably continued in China during a long period while sediments were contemporaneously deposited. Pumpelly was the first to direct attention to these wide-spread events. But it is only in the great granitic region of the eastern coast, between Ning-po and Hong-kong, that porphyry itself arrives at an extraordinary development. The Chusan islands are almost exclusively composed of quartzose porphyry and its tufas, and from there southward it appears to be only subordinate in quantity to the granite. I know it from my own observations on the island of Hong-kong, and by inference from the observations of others of the region between that island and Ning-po. This is the most extensive development of porphyry known in any part of the world.

Deposits of porphyritic tufa, sandstones, and clays.—The porphyries themselves are little developed on the lower Yang-tze. I noticed their first appearance in certain porphyritic tufas which overlie somewhat unconformably the Sangbu sandstone. The latter appears, indeed, from its purely siliceous character, to have been antecedent to any outbreak of porphyry, while the soft and impure nature of all subsequent deposits goes to show that they were the tufaceous sediments of eruptions in remote regions. The visible thickness of this formation below Hankau is about 3,500 feet. It incloses a few beds of coal of subordinate value.

Herewith ends on the Yang-tze the series of ancient formations.

It will be observed that von Richthofen gives no clue to the age of these deposits, which, in view of the unconformity at their base, probably do not include the earlier transition sediments, but may comprise the Triassic and

---

*China, vol. II, p. 368.
† American Academy Arts and Sciences, vol. VIII, 1869, p. 117.

also the Lower Jurassic. The Sang-hu is thus probably nearly equivalent to the Kui-ehóu series.

On the Yang-tzï still, but several hundred miles further west, occur the exposures of Permo-Mesozoic strata which form the foothills along the mountains of western Hu-peï. Pumpelly described them as he saw them near I-chang, and his account is given in volume 1, p. 278, of this report. Vogelsang* repeatedly refers to them as being observed in a trip from I-chang northward as far as the 32d parallel, from which he returned via Chu-shan and the Ta-ning-ho; the latter part of his route coincides with that subsequently followed by our expedition.

We observed strata above the Upper Carboniferous (Wu-shan) limestone in numerous sections on the Ta-ning-ho and Yang-tzï-kiang, but did not examine them closely. There was great uniformity of stratification and, despite the marked difference between the massive limestone and the red shales which followed it, seemingly perfect conformity of dip between the two. Both in the field and since, I have regarded the relations as those which result from marine scour by shallowing waters. The limestone, having accumulated and hardened beneath waters which were deep enough to permit sediments to gather, was swept clean by a current which flowed across it, possibly with accelerated velocity, while the waters shallowed; and ultimately, in consequence of continued shallowing, the current was checked or diverted and red terrigenous deposits were laid down on the scoured surface. This sequence implies the nearness of land areas and possibly of subaerial erosion of the Wu-shan limestone in adjacent areas to a degree consistent with the occurrence of such conglomerate of flints as we observed in the float of the Ta-ning-ho.†

In the limestone, 400 feet, 120 meters, above the base of the red series, occur fossils which Girty refers to the late Paleozoic, but with doubt.‡ He says:

> The evidence upon which the horizon in question is referred to the Paleozoic consists of the septiferous Terebratuloid, the round crinoid stems, and the general Paleozoic complexion of the meager fauna. It is inconclusive and may not stand against any facts which strongly point to a Mesozoic age. For this reason my assignment to the Carboniferous is provisional merely, and should the necessity arise of changing to a younger period, the readjustment would entail no serious conflict of evidence.

Girty's conclusion from the fossils is in accord with the stratigraphic evidence that deposition was more nearly continuous here than in areas adjacent on the north and west. The episode of marine scour, which is

---

* Reisen in nördlichen und mittleren China, Petermanns Mitt., 1901, 47, and 1904, 50.
† Vol. 1, pp. 264 and 277.
‡ Report on Upper Paleozoic Fossils, vol. 111.

supposed to account for the sharp contact of the red shale on the Wu-shan limestone, may be assumed to be of less duration than that represented by marked unconformity of dip; and therefore the strata succeeding upon the area of marine scour would probably be older than those transgressing upon the area of subaerial erosion.   The former might be late Paleozoic, as Girty suggests.   The latter probably range from early Mesozoic to Jurassic.

The middle Yang-tzï region, where the strata have been observed by Pumpelly, von Richthofen, and ourselves, is continuous on the west with the Red Basin of Ssï-ch'uan.   The next notable observations on the Paleozoic-Mesozoic rocks relate to the northern margin of that basin near Kuan-yüan-hién, in the region reached by von Richthofen, Lóczy, and Obrutchov, and already several times referred to.

South of the section exposed by the Kia-ling-kiang across the folded Paleozoics are strata which dip gently southward into the Red Basin.   There is a marked unconformity, above which the first formation is a massive limestone.   Von Richthofen regarded it as probably Permian or Triassic.* Lóczy describes the section observed by von Richthofen in nearly the same terms,† and refers the limestone, which has a thickness of 1,400 feet, 400 meters, or more, likewise to Permian or Triassic.   He, however, leaves the possibility of a still younger age (Rhætic) open.   In certain thin-bedded, light-colored, marly limestones, which are the uppermost strata of the folded series beneath the undisturbed supposed Permian limestone, he collected indistinct fossils, on which he comments as follows:‡

> Among the materials which I collected there occur a cast which resembles *Megalo-donta*, an *Aviculopecten*, as well as numerous examples of that little bivalve which we commonly designate by the indefinite name of Myacites.   These forms permit us to suppose that the marly clays and thin-bedded limestones which overlie the supposed Carboniferous sandstone may be of Upper Carboniferous age, and that they perhaps are the southern representatives of the North China supra-Carboniferous series.   But the possibility is not excluded that even Permian and Triassic may be included among these disturbed strata, and that the evenly bedded limestone which occurs at the base of the middle Jurassic sandstones, in the Basin of Ssï-ch'uan, may belong to the Rhætic system.   I found, in fact, the traces of this system in the western margin of the Basin of Ssï-ch'uan, as we shall see further on.

The observation to which Lóczy refers in the preceding paragraph is probably that which he records on pages 736 to 739.   In an excursion in the vicinity of Tschung-tjen he observed Triassic strata which were apparently conformably bedded between two heavy limestones, both of which he assigns to the Carboniferous.   He explains the relations by a conformable

---

*China, vol. II, p. 603.   Quoted in full in this report, vol. I, p. 295.
† Reise des Grafen Széchenyi, vol. I, pp. 439–440.
‡ *Ibid.*, vol. I, p. 441.

contact at the base and an overthrust above. The Triassic strata consist of sandstones and fine-grained sandy shales, which contain several species of Myophoria, as well as impressions of *Encrinus liliiformis*, together with other well-preserved forms. The locality is in western Yün-nan, latitude 28°, longitude 100° east.

The Red Basin of Ssi-ch'uan was first explored by von Richthofen, but his observations still remain unpublished. Lóczy crossed the northwestern margin only, yet his account affords the best available information.[*] It is quoted in the section on the Jurassic.

The occurrences which have been described for the Yang-tzi and Ssi-ch'uan are those of strata south of the metamorphic district of the Han valley, which borders the southern Ts'in-ling-shan. In the vicinity of Shi-ts'üan-hién, along the middle course of the Han, and in that part of southeastern Shen-si which lies between Hing-an-fu and Chōu-p'ing-hién, the expedition of 1903–04 observed certain schists, which appear to hold the stratigraphic position of the Kui-chóu series and to have the mineralogical character that metamorphic rocks derived from those sediments should have. They have been described in volume I, chapter XIV.

We observed these metamorphosed Fermo-Mesozoic strata about Shi-ts'üan-hién in latitude 33°, longitude 108° 30' east. The next occurrence toward the west which may be assigned to this horizon is described by Lóczy,[†] and presents peculiar characters. It lies north of Kuan-yüan-hién and is involved in the folds of the Paleozoics near Chau-tién. The strata overlie the typical Carboniferous limestone and consist of slaty limestones and sheared clay slates, which are reddish-brown in color and inclose large limestone blocks. Lóczy suggests that they are of Silurian age and are overthrust on the Carboniferous, and this may be the case. But it seems equally probable that the beds are Permo-Mesozoic and differ from those of the same age near Kuan-yüan-hién simply in degree of slatiness. If so, the relations are similar to those which exist between the altered rocks north of Chōn-p'ing-hién, Shen-si, and the unaltered beds south of that town. The inclusion of large limestone blocks is not explained. They may be lenses, peculiar conglomerates, or of autoclastic nature.

JURASSIC.

Jurassic strata occur extensively in central and southern China, and also in northeastern China in the provinces of Shan-tung and Liau-tung. The basin of Jurassic coal-bearing strata near Ta-tung-fu in northern Shan-si, longitude 113° east, latitude 40°,[‡] is apparently an isolated occurrence

---

[*] Reise des Grafen Szécheyi, vol. I, p. 685.
[†] Ibid., vol. I, p. 438.
[‡] China, vol. II, pp. 356–359.

far to the northwest, beyond which in Mongolia no deposits of similar age occur.* Following the trend of the mountain ranges from Ta-tung-fu southwest through Shan-si, we come upon the Jurassic in the eastern Ts'in-ling-shan† whence it may be traced through various synclines westward across Shen-si to the Red Basin of Ssī-ch'uan. The line thus followed constitutes the northern limit beyond which neither Lóczy nor Obrutchov report any observations of Jurassic deposits.

The formations of this period in China are of the continental type. They consist chiefly of sandstones, with shale, generally red, more or less associated with coal, devoid of marine fossils, but characterized by plants which, north of Indo-China, do not contain Gondwana species, but are frequently related to Russian Jurassic forms.

I proceed to the enumeration of some characteristic occurrences.

In Shan-tung, in the coal-basin of Weī-hién, are strata from which Lorenz collected Jurassic plants.‡

In the eastern Ts'in-ling-shan, near Shan-chóu in Shen-si, Lóczy observed a basin of Jurassic strata which rest on ancient metamorphic schists. He says:§

Back of Shan-chóu the valley is bounded by steep cliffs, in which the steeply up-turned transgression of the basin deposits is clearly exposed. As is shown in fig. 30 a, the conglomerate beds which, in a gently inclined attitude, underlie Shan-chó u, assume a constantly steeper, almost vertical dip; they are underlain by fine-grained clayey sand-stones alternating with dark marly layers, and finally there come in dark, bituminous, thin-bedded marls, with thin sandstone layers. These last constitute the lowest strata in the basin, and are much folded as a whole in a width of a kilometer. In the marl as well as in the sandstones occur numerous carbonized fruits (carpolithus) which Schenck described as belonging to cycads or as coniferous fruits resembling taxinea. According to him they resemble the fruits of Bairia or Gingko of the Jurassic of Siberia.

The occurrence thus described by Lóczy is peculiar, in that it lies so far in the heart of the Ts'in-ling-shan and the strata are more strikingly folded than is common with those of like age.

Some miles south of the above-described occurrence, between "Tse-chuen" and "Kiu-tze-kuan," Lóczy crossed an extensive basin filled with deposits which lie in horizontal attitude and consist of the local sediments from the surrounding mountains. He speaks of them as Mesozoic, but does not cite any evidence, and it would appear not improbable that they are of later age.

*Lóczy: Reise des Grafen Széchenyi, vol. II, p. 799.
† Ibid, vol. II, p. 415.
‡ Beitrage zur Geologie Shan-tungs, Heft I.
§ Reise des Grafen Széchenyi, vol. I, p. 415 fig. 30 a.

Strata which we provisionally assign to the Jurassic occur at Shï-ts'üan-hién on the Han river, in the form of red and yellow sandstone and conglomerate composed of the detritus from older rocks of the vicinity. The area is apparently limited and owes its present position and preservation from erosion to a fault by which the soft young strata are brought to a relatively low level among the metamorphosed Paleozoics.*

Still farther west, in the northern margin of the Red Basin, are the Jurassic beds near Kuan-yüan-hién, which were first recognized by von Richthofen†, again observed by Lóczy,‡ and repeatedly referred to in this work. Von Richthofen's section is quoted in volume I, page 295. Above the coal-bearing middle Jurassic, as determined by plant remains, come coarse conglomerates, 300 feet, 90 meters, thick, followed by yellow, reddish, and finally greenish sandstones about 200 feet, 60 meters, thick. The conglomerates appear to mark an invasion of coarse material of local origin, which may be ascribed to increasing declivities or to greater precipitation in nearby mountains. Together with the overlying sandstone it may with reason he referred to the middle or upper Jurassic.

This occurrence is presumably typical for the Red Basin of Ssï-ch'uan, so far as its margins are concerned. Although von Richthofen made more extensive observations in this great interior depression of western China than any one since, we have no account of them and rely upon Lóczy, who traversed the northwestern margin. He states:§

Immediately upon the early Paleozoic deposits, on the margin of the basin, follow the Mesozoic strata. In the vicinity of Ja-chóu-fu the Carboniferous appears to be wanting, as immediately beneath the Ssï-ch'uan sandstone we found Devonian limestone near Hoa-ling-pu. Near Kuan-juön, on the other hand, the Carboniferous and Permian strata probably occur in the marginal mountains, as has been stated above.

The Mesozoic systems are represented by the great series of the Ssï-ch'uan sandstones. This sandstone formation consists of red and gray sandstones, coarse conglomerates, and marly shales, and in its lower layers are huge coal-beds, which are worked at many points about the northern and western margins. The coal-beds near Kuan-juöu-shien and Hoani-pu yielded impressions of plants of the middle Jura (Dogger), whereas near Lin-tschi-shien and Ni-tou we collected Liassic plants.

Marine fossils are not yet known from the Ssï-ch'uan sandstone. The only fragment of a mollusk is a doubtful *Anoplophora* (Cardinia) which was found near Lin-tschi-shien, but it is not sufficiently well preserved to afford definite evidence for or against a marine origin.

The Ssï-ch'uan red series is petrographically as well as geologically and structurally identical with the red sandstone of the province of Kiang-si, as well as with the basin deposits which we saw on the lower Han-kiang and on the Sié-ho.

---

* Research in China, vol. I, p. 316
† China, vol. II, p. 603.
‡ Reise des Grafen Széchenyi, vol. I, p. 440.
§ *Ibid*, pp. 685–686.

PERMO-MESOZOIC DIASTROPHISM.

The diastrophic movements which occurred in all continents during the closing epochs of the Paleozoic and the initial epochs of the Mesozoic were pronounced and prolonged throughout Asia. They ultimately changed that face of the globe, as Suess has pointed out,[*] welding together the separate elements of the Asiatic continent, except that the Angara and Gondwana lands remained separated by the Himalayan strait. In contrast to the gradual changes of level which had characterized the Paleozoic, these disturbances were of decidedly orogenic character. They gave rise to mountain systems, which are structurally still the controlling features of Asia. The foundations of the ranges are now raised to the summits of the Tién-shan, Kuen-lung, and Ts'in-ling-shan, and the substance of their masses constitutes the Triassic and Jurassic sediment of Asia. By Cretaceous time the continent was again low.

The continental structure of Asia has been so elaborately described by Suess,[†] the master of the subject, that extended discussion is superfluous unless based on new data; and in general such data are not yet available. Furthermore, what later information we have confirmed his deductions, as regards the trends of mountain axes. We may, however, point out that the Sinian is more generally folded in North China than von Richthofen, and consequently Suess, supposed. And our observations give additional data on the structure of the Wu-t'ai-shan, the Ts'in-ling-shan, the Han district, and the middle Yang-tzï region.

On the map, plate 8, are shown the structural lines given by Suess,[‡] supplemented by those traced by Futterer[§] south of latitude 36°.

Referring to these authors for their presentation, I proceed to discuss some of the newer significant facts.

Through his reconnaissance observations in Shan-tung von Richthofen gained the impression that Sinian strata had not been folded, and in passing through the Wu-t'ai district he was so circumstanced that he did not observe the actual facts of structure. Our observations in the Ch'ang-hia district of western Shan-tung showed that the Sinian is folded and even overthrust, the axial trends being dominantly northeast-southwest and subordinately northwest-southeast.[||] And in the Wu-t'ai district the Sinian and Carboniferous exhibit overfolds and overthrusts of a decided

---

[*] La Face de la Terre, vol. III, p. 24.
[†] Ibid., vol. III.
[‡] Ibid., vol. III, pl. III.
[§] Petermanns Mitt. Erg., Heft 119.
[||] Vol. I, chapter III, and plates XIII and XV.

character, which there extend northeast to southwest, parallel to the structure of the pre-Sinian sediments.*

To what extent these structures are general in North China is not yet determinable. The Ch'ang-hia district is a very limited area; the folding is moderate and probably local. The folds of the Wu-t'ai district are much more pronounced and occur in a zone of ancient deformation which is marginal to the masses of Mongolia and Ordos. It is traced from northern Chï-li across Shan-si to Shen-si, a distance of 500 miles, 800 kilometers, and is a structural zone of the first order, which is represented in modern mountain systems by notable heights. Southeast of it lies the plateau of eastern Shan-si, which, in the southern part where von Richthofen crossed, presents the Paleozoic in little disturbed strata. It is possible, but not known, that no folds invade its broad area.

The date of deformation in North China is fixed as not earlier than upper Carboniferous (Pennsylvanian) by the parallelism of strata of that age with the Sinian over wide areas, and as pre-Liassic by the unconformity which was observed by von Richthofen between the Sinian and the Liassic coal-measures of the Ta-tung-fu field in northwest Shan-si. The latter strata are also tilted, probably in consequence of a recurrent movement in Jurassic time.

As was first pointed out by von Richthofen, the Ts'in-ling-shan, the eastern continuation of the Kuen-lung system, occupies the site of a trough in which Paleozoic sediments accumulated to considerable thickness, if the consensus of opinion regarding the age of the metamorphosed sediments be correct. Its folds embrace the southern margin of the ancient land mass that is now northern Shen-si and Kan-su. The trend of the axes of folding in that part of the range east of longitude 108° is not exactly parallel with the modern height. Von Richthofen's preliminary map, a bold essay on the slight data available to him, was mistaken in this respect. The range trends about north 75° east, and the general axial strike of the structures varies from east-west to north 80° west. Thus the belts of the Archean, Proterozoic, and Paleozoic strata in general approach the northern front at an angle and are cut off by the fault, which is the latest tectonic feature. The angle is an acute one, and in the case of the contact of the Proterozoic and Paleozoic south of Chóu-chï-hién, the line trends south of west; but the larger features of the structure diverge from the range and pass under the Weï valley, where they are faulted down.

The Ts'in-ling-shan and the mountainous region south of it, through which the Han flows from Han-chung-fu to beyond Hing-an-fu, was the scene of more or less intense metamorphism and intrusion, as well as of

---

*Vol. 1, chap. v.

folding.   In the section observed by Lóczy, in longitude 106° east, in that
previously observed by von Richthofen in longitude 107° 30′ east, and in
that of the expedition of 1903–04 in longitude 108° 15′ east, there are areas
of slates, argillites, schists, and gneisses, apparently derived from Paleozoic
sediments and associated with large masses of intrusive diorite or granite.
In the last-named section two large granite masses and several smaller ones
present a total width of 19 miles in 84 miles.   They are intrusive bands,
whose length along their trend parallel to the axis of the range is very
considerable.   Their effect in altering adjacent strata is extremely variable,
the limestone of the Heï-shui series near Liu-yüé-ho (atlas sheet a 2) being
but little affected a hundred feet from the contact, whereas near Ssï-móu-ti
(atlas sheet a 3) the Paleozoic strata are generally and intensely metamorphosed, although the intrusions as they appear at the surface are much
smaller.

The date of these intrusions is later than the folding, as the holocrys-
talline intrusive rocks were not sheared, but it is presumably not much
later.   Among the intruded and metamorphosed strata we believe we have
identified the K'ui-chóu formation, i. e., Permo-Triassic, which may have
been contemporaneous with the early effects of deformation, but which was
involved in the later effects to the extent of overfolding and overthrusting.
And unconformably above the metamorphosed strata occur the unaltered
Shï-ts'üan sandstones, which we correlate tentatively with the middle or
upper Jurassic sandstone of the Red Basin of Ssï-ch'uan.   These data seem
to confine the episode of intrusion to the Triassic or early Jurassic, to the
close of the period of diastrophism.   The inference is strengthened by the
fact that there are not any Cretaceous or early Tertiary sediments of marine
or continental character, such as should occur as a result of erosion if,
during those periods, there had been great intrusions of granite, with the
probable accompaniment of decided elevation.

The region within which the early Mesozoic intrusions occur is not yet
well defined.   In the western Ts'in-ling-shan, between longitudes 107° and
108° east, the zone may be said to extend from the Weï to the Han valleys,
between latitudes 33° and 34° 10′ north.   In longitude 109° 30′ its southern
margin is near Chöu-p'ing-hién, latitude 31° 50′.   The eastern and northeastern extension is indefinite, as the great granite masses of the eastern
Ts'in-ling-shan, which may in part belong to this period, are described
by Lóczy and von Richthofen as "Archean."   Toward the northwest, in
longitude 105° east, latitude 35°, near "Kun-tschang-fu" Lóczy * observed
intrusions of granite in mica, amphibole, and chlorite schists.   The occurrence is one of many granite masses which characterize the northwestern

* Reise des Grafen Széchenyi, vol. I, p. 425.

Ts'in-ling-shan, and appear to be part of the zone under consideration. The western continuation of the mountain system, the Nan-shan ranges, was a scene of igneous intrusions, as described by Obrutchov, but whether of pre-Cambrian, middle Paleozoic, or Mesozoic date we can not distinguish. In western Ssï-ch'uan, longitude 99° to 103° east, latitude 30° north, between Ba-tang and Ja-chóu (Ya-chóu), where Paleozoic and Mesozoic strata are folded around the eastern margin of the Tibetan plateau, Lóczy observed great bodies of granite in the metamorphic schists. He compares the rocks with the Nan-shan sandstone, and again we are uncertain whether the intrusions are pre-Cambrian or Mesozoic. In this connection it is worth while to point out that the Ts'in-ling-shan and the Alps of western Tibet bound the depressed basin of Ssï-ch'uan on the north and west and coalesce at its northwestern corner. The geologic conditions which led up to the Fermo-Mesozoic diastrophism were identical in the two belts, so far as the stratigraphic history is concerned; the effects of folding appear to have been much the same; and it is probable that the phenomena of intrusion which affected the one occurred in the other. The southern continuation of the mountains of Ssï-ch'uan and Yün-nan extends through Indo-China, where Fuchs and Saladin distinguish an ancient granite and a microgranulite of Carboniferous or post-Carboniferous age.[*] In view of the reconnaissance nature of the observations the age of the ancient granite may be considered undetermined and possibly Mesozoic.

In this connection it is desirable to mention the fact brought out by Hayden[†] in course of the Younghusband expedition to Tibet that Jurassic and Cretaceous strata of southeastern Tibet, south of Lhasa, are highly metamorphosed and intruded by granite. At first thought the suggestion occurs that the post-Cretaceous age of these intrusions may indicate a like recency for the intrusions of the Ts'in-ling-shan; but the unaltered character of the supposed Jurassic in that region appears to preclude the idea; and it is more likely that the events in the Ts'in-ling-shan and in southeastern Tibet are successive than that they were contemporaneous.

In strong contrast to the pronounced folding, metamorphism, and intrusion that characterize the central Ts'in-ling-shan and the Han valley is the moderate deformation of the region on the south. Folded Paleozoic strata, which are indeed closely folded and even overthrust, but not altered, adjoin the metamorphic district along a remarkably sharp boundary, and pass into the wide flexures of the middle Yang-tzï or the synclinorium of the Red Basin. On the Yang-tzï there is complete conformity of dip up to and including the K'ui-chóu beds (Rhætic). About the Red Basin there

---

[*] Annales des Mines, 1882, 8me série, Mémoires 2, pp. 224–225.
[†] Report Geol. Sur. of India, 1905.

are unconformities. That near Kuan-yüan-hién is overlain by limestone which may be Permian or Rhætic, as already stated. But west of Ja-chóu* the Mesozoic sediments are conformable to the underlying Devonian and are folded with it. Hence Lóczy infers that the folding continued longer in the latter region. It is possible that the vicinity of Ja-chóu was outside the zone of Permo-Mesozoic folding, as most of the Red Basin was, and became involved in disturbance only when the Himalayas were compressed in the great post-Eocene epoch of diastrophism. In the zone of the great mountain chains of northern India, the Permo-Mesozoic movement was manifested only in variations of sediment, not in folding.

---

*Reise des Grafen Széchenyi, Lóczy, vol. I, p. 690.

CHAPTER VII.—CONTINENTAL ASIA.

CRETACEOUS.

A geologist taught only by observation in China, outside of Tibet, would know nothing of the Cretaceous. No strata are known which may be correlated with the strata that represent the period in America and Europe. We would look for marine or continental deposits in the middle Yang-tzï region or the Red Basin of Ssï-ch'uan, where they might cap the Jurassic in the deeper synclines; or about the margins of the great alluvial plain of the Yang-tzï and the Huang-ho, where they might outcrop in the foothills; but they are not found. If they ever existed above the known Jurassic they have been eroded; if they underlie the plain they are overlapped.

Nor are Cretaceous strata of any kind known in the vast area of Asia north of Tibet, east of the Urals, and south of northern Siberia. No other fact than this, perhaps, more sharply challenges the hypothesis that the present mountain ranges and basins of central Asia date from a pre-Cretaceous time. Highlands without waste and waste without deposit in these interior basins are inconceivable; but though the plateaus are affected by post-Cretaceous dislocations which expose even the ancient crystallines, there is no trace of deposition during the Cretaceous period. The surface that could for so long a time maintain a blank record must have been a land in which transportation and aggradation had ceased in consequence of uniformity and flatness of slope—a well-developed peneplain. Even on such a surface subaerial decay must produce residual material and the atmosphere do some work; but the product might be worked over into later deposits, as it probably has been.

Some of the waste from this Cretaceous land found its way into the southern Tethys, where it formed the Giumal sandstone. The area of deposition extends throughout the Himalayas, over southern Tibet, and through Kashmir into Afghanistan and Persia.*

The rock is a greenish-gray sandstone, sometimes very siliceous, and of considerable thickness. There is a transition to it from the underlying Jurassic shales, and it is separated by a sharp but conformable contact from the overlying upper Cretaceous limestone, which extensively overlaps it. Thus, in the latest Jurassic, lower Cretaceous, and upper Cretaceous of

---

*Griesbach: Memoirs G. S. I. xxiii, p. 81.

northern India we have the sedimentary record of the reduction, peneplanation, and partial submergence of the continent, which in preceding Mesozoic
time had attained very prominent relief.

TERTIARY AND QUATERNARY.

A sedimentary record of the Tertiary history of China is wanting, as
is that of the Cretaceous.   We turn to Indo-China, India, southern Tibet,
and western Asia, for occurrences of marine and estuarine deposits, which
by their calcareous or carbonaceous character show that the peneplain
conditions of the Cretaceous period were continued, at least in southern
Asia, through the Eocene and Oligocene and well into the Miocene.   The
marine zone which was the southern branch of the Tethys persisted along
the Himalayan region and still divided the ancient Gondwana land of the
Indian peninsula from Tibet.   The movements which, during the Oligocene
and Miocene, closed the strait, are described by Griesbach,* who, in a limited exploration of the Hundes plateau of southern Tibet, distinguished
altered nummulitic limestone, unconformably overlain by Miocene (?) sandstone, a few hundred feet thick, which in turn is conformably covered by
a great thickness of nearly horizontal beds that Lydekker determined on
the evidence of mammalian bones to be probably Pleistocene and certainly
not older than Pliocene.   The recent observations by Hayden in the province of Tsang and U along Younghusband's route to Lhasa have confirmed
Griesbach's observations.†

The evidence of stratigraphy north of the range agrees with that of
the Siwaliks of the southern foothills, and there is every reason to accept
Oldham's view that much of the elevation of the Himalayas has occurred
since the Miocene epoch.‡   In view of the evidence that other mountain
systems of Asia have grown to great heights during the Pleistocene it may
be questioned whether the Himalayas reached their present altitude during
the initial movements of folding; it is even probable that they have suffered
one partial epoch of erosion to advanced maturity and have since been
warped up, as have the Apennines and Karpathians;§ but they in any
case represent the latest effects of the compressive force which has welded
Asia into a continent.

History repeats itself, now here, now there.   The northern Tethys
was closed by a late Carboniferous movement, during which the Kuen-lung
system of folds developed; and the supra-Carboniferous sandstones were

---

*Central Himalayas, Memoirs G. S. I. xxiii, pp. 82–87.
†Geology of Tibet, Rec. G S. I., 1905.
‡Manual of Geology of India, second edition, 1893, p. 479.
§Studies in Europe, B. Willis, Year Book of the Carnegie Institution of Washington No. 4.

spread upon the flanks. The strata that form the Ts'in-ling-shan and the Alps of western Tibet shared in the Kuen-lung movement, but they also suffered a later intense disturbance, which gave rise to the erosion cycle that is represented in the Jurassic sandstones. And the Himalayan zone, which, until post-Eocene time had not been folded, then became the locus of pronounced deformation and passed through the movements that are evidenced by the unconformities and sediments of the Tertiaries. Lóczy noted that the Jurassic beds of Ssï-ch'uan and the supra-Carboniferous of central Asia* are similar products of vigorous erosion of mountain ranges that were elevated at separate times; and we may add the Hundes and Siwalik sandstones as a third group of formations of the same kind.

With the epoch of mid-Tertiary folding the compression of Asia ceased for a time at least. The flat Hundes sandstone has its equivalents in the essentially flat Pliocene and Pleistocene deposits of central Asia and Siberia. The record is one of erosion and wide-spread deposition in basins, either lacustrine or arid, and on fluviatile plains. But diastrophic movements have not ceased; they have taken on the form of normal faulting, involved in spreading of the continent, and the major features are accented by the fractures. These phenomena, which are extreme effects of vertical warping, are chiefly of Pleistocene age.

I proceed to consider the evidence of orogenic movements other than folding; it is subordinately stratigraphic and predominantly physiographic. Before considering the Quaternary warping it is well to describe in proper sequence the Tertiary dislocations of the same type. In northeastern China, in the provinces of Shan-tung and Chï-li, are two districts in which normal faulting is the principal structural fact. They have been mapped and described in volume I, so far as we saw them; and we have there cited observations by von Richthofen, who noted the faults of Shan-tung as the principal structural facts of that province. The district in Chï-li is that which we call the Ning-shan basin, west of Pau-ting-fu, and has not been seen by any other geologists.

The evidence of faulting in these cases is chiefly stratigraphic: Paleozoic faulted down in contact with Archean, along throws which range up to 10,000 feet, 3,000 meters, or more; but in Shan-tung the relief due to displacement still survives, though it is greatly dissected. Fault-scarps have receded one or two miles; fault-blocks are cut into isolated sections by the valleys of consequent streams that originated on the scarps; valleys have widened and mountain masses have become skeletonized. The volume of rock removed is very large, but since the districts have had open drainage to areas now submerged or buried beneath recent alluvium, there are no

* Reise des Grafen Széchenyi, vol. I, p. 799.

accessible deposits corresponding to the cycle of erosion. We are thus restricted to physiographic evidence in seeking to date the epoch of faulting, but, by comparing the effects of erosion with those accomplished elsewhere since the middle Tertiary, we reach a probable conclusion that the time was pre-Miocene.

In the Ning-shan district of Chï-li the relief due to faulting no longer survives; indeed it has been reversed, limestone hills on the downthrow rising 1,000 feet, 300 meters, above the surface of gneiss on the upthrow. Hence we infer that the faulting may be somewhat older than that in Shan-tung, and we place it in the earliest Tertiary.

The faulted district of western Shan-tung and that of Ning-shan in Chï-li are isolated occurrences in eastern Asia, so far as we now know, of dislocations of early Tertiary age. Yet it is probable that they are not singular, and anticipating somewhat the discussion of physiographic cycles which follows, it is desirable to state in this connection the conclusion reached by Suess, that the profound graben which is occupied by Lake Bai-kal has existed since the close of the Tertiary period.* The evidence, which consists of the survival of species of European Pliocene affinities in the lake, is clearly assembled, and the age of the basin appears well established. The depression is due to normal faults, which define the graben and which are paralleled by others that give rise to the ranges of Trans-Baikalia. Attrib-uting to the system the age determined for the lake basin, we recognize in Trans-Baikalia a mountain group of late Tertiary date.

It will be seen by reference to volume I, that the faults which charac-terize the Ki-chóu-shan, Ho-shan, Hua-shan, and Ts'in-ling-shan, all of which are mountain ranges in Shan-si and Shen-si, China, are referred to a Quaternary, probably middle and late Quaternary, date. Hence, taking account of the Chï-li, Shan-tung, Baikal, and Shan-si fault-systems, I con-clude that normal faulting has been a feature of orogenic activity in Asia, in one district or another, since early Tertiary time.

Warping, that is, nearly vertical displacement of different parts of the surface to unequal amounts and often in opposite directions without dislo-cation, has been a general effect of diastrophism, especially during the later Tertiary and Quaternary. And the displacements have been so conditioned in time and place as to give rise to cycles of erosion, which can be distin-guished in the plains, plateaus, ranges, and rivers of the continent. They have been described in the physiographic study of the districts through which we passed, as presented in volume I. There the surface is analyzed, the development of streams is traced, and the interaction of diastrophic

---

* La Face de la Terre, vol. III, p. 78.

movements with erosion is presented.   The climatic factor is also recognized and applied to aid in fixing the date of initial loess deposition.

Four phases are distinguished: the first or oldest is a peneplain, a very ancient and also very aged form, which is known from various parts of northern Asia, and a remnant of which we named from its preservation in the highest dome of the Wu-t'ai-shan, the Peï-t'ai form, developed during the Peï-t'ai cycle.

The next younger is a surface of mature erosion, which replaced any older features in most of the areas we saw.   It is a surface of moderate relief, characterized by wide valleys and hills rarely a thousand feet high. It is typically developed near T'ang-hién, Chï-li, and we call it the T'ang-hién stage.

The third phase was one of aggradation in North China, the time of the early loess deposits.   The moderate relief of the preceding stage was to a notable extent buried beneath the Huang-t'u, a formation consisting of wind-sorted waste from the deserts of central Asia, whence the dust was brought chiefly by rivers.   The partly buried hills along the western margin of the Great Plain of eastern China afford an illustration of the aspect of Chï-li, Shan-si, and northern Shen-si at the time.   The great mountain ranges had not attained their present height.   Attributing the desert waste to the climatic change from Tertiary to Pleistocene, which may have become effective in late Pliocene to the extent observed, we assign this phase to that time and to the opening of the Pleistocene.   We designate it the Hin-chóu stage, after the Hin-chóu loess basin in Shan-si.

The fourth and present physiographic stage we named for North China the Fön-ho, from the river of that name, which, though older than the Fön-ho epoch, still flows through Shan-si among characteristic features of that stage.   For South China, where the physiographic relations are somewhat different, we applied the name Yang-tzï to what is very nearly or precisely the same time division.   It is an epoch of very decided mountain growth in China; and if, as I believe, the principal continental upwarp of central Asia is largely of the same date, it is the time of one of the most remarkable diastrophic movements of which we have knowledge.   It appears to fall chiefly within the Quaternary, but may extend back into the Pliocene. The typical features are warped and faulted surfaces, which result from downward and upward movements of adjacent masses that underlie basins and graben or constitute plateaus and mountain ranges.   The amount of sculpture is relatively slight, but great canyons like the Yang-tzï gorges have been cut by antecedent rivers.

Having thus summarized the results of our observations in China, I suggest their broader relations.

The oldest, the Peï-t'ai cycle, links the physiographic history with the stratigraphic record. We have seen that in the sediments there is reason to regard the Cretaceous as a time when Asia presented the aspect of an extensive peneplain, over which the sea transgressed from the south. Sediments of the Eocene and Oligocene epochs record the same conditions. We regard the Peï-t'ai as the equivalent surface, and the Peï-t'ai cycle as covering late Cretaceous and early Tertiary. It is the Asiatic representative of the Schooley or Kittatinny peneplain of the eastern United States and of the Cretaceous peneplain of central Europe.

The remnant of a peneplain which we recognized in Peï-t'ai by its form and by the residual soil peculiar to it, is of very small extent, and we may well ask what other evidences remain to support the inferences of a once general condition. Let us proceed northeastward from the Wu-t'ai-shan, which is in northern Shan-si, and review some known features, bearing in mind that the ancient surface is both warped and eroded; consequently it may occur at any altitude and may be more or less dissected. Where the Siberian railroad, after leaving the dislocated mountainous region of Trans-Baikalia, traverses northeastern Mongolia to the Khingan range and descends into the valleys of Manchuria, there is a plateau surface which is a slightly warped plain of erosion, occasionally capped by lava flows. In the Khingan range it is warped a few hundred feet higher and, extending over the crest, is represented in the summits of the long spurs which constitute the deeply canyoned eastern slope. In Manchuria it sinks beneath the alluvium of the Sungari, as the tilted peneplain that forms the western slope of the Sierra Nevada of California sinks beneath the alluvium of the Sacramento.

Suess refers to the eastern slope of the great Khingan range in its northern extension as being analogous to a flexure, and gives the following account of the great plain of the Amur:*

The constitution of the plain which extends northward from the upper Amur is not altogether simple. Even at the eastern base of the Khingan, beds of the Angara series appear * * * and occupy all the west of the plain * * *. They are covered by white sands and shales with lignite of Tertiary age, which occupy the eastern portion of the plain as far as the Zéia river * * *. But this mantle of Tertiary has in general only a slight thickness; beneath it appear, as upon the upper Tygda, the Archean rocks which play scarcely any orographic rôle. * * *. In the occasional outcrops of the Archean basement M. Ivanov recognized the general strike of north-northeast, but according to this observer the Angara strata which form the western part of the plain are not horizontal; on the contrary, along the Amur, in the strip between the great Khingan and the first outcrops of the Archean basement, they appear folded with the same north-northeast strike. The plain owes its origin to the degradation of these folds.

*La Face de la Terre, vol. III, p. 155.

This description gives definite information regarding the occurrence of a peneplain which is clearly older than the Tertiary deposits which cover it, and may fairly be correlated with the Peï-t'ai plain.

With reference to the mountains which lie north of the Amur between longitude 120° and 140° east, and which are known as the Aldan or Stanovoi range, Suess quotes Krapotkin as saying, "That the supposed continuous chain of the Stanovoi range, serving to divide the waters of the Arctic ocean from those of the Pacific, does not exist, whether one thinks of it as high or low, abrupt or flattened."*

And again, referring to the Vitim plateau, which is north of the ranges of Trans-Baikalia, between 110° and 120°, he quotes Krapotkin to the effect that over a great distance the country has lost all individuality.

The travelers who have there sought to follow the line dividing the two slopes have not discovered a long and continuous mountain chain, but have found, on the contrary, virgin forest, rocks covered with moss, and vast swamps interrupted here and there by lakes.

The region thus described is connected by a long slope with the plain of the Amur, and descends on the northwest to the great plateau of horizontal Paleozoic rocks in which the Lena has sunk its modern canyon. There can be little doubt that the vast expanse of plateau and plain in northern Siberia, from the range of the Verkojansk mountains on the east to the recent alluvium of the Ob on the west, is a great peneplain like that of the Canadian highlands of North America; and traced as it is by the unbiased observations of Krapotkin, Ivanov, and other observers, to a position beneath the Tertiaries of the Amur basin, it may well be assigned to an early Tertiary and Cretaceous epoch.

I have seen the representative of this peneplain in the vicinity of Krasnoyarsk, where it forms the summit of the hills that bound the valley of the Jenisseï, and observed it in the foothills of the northern Altai as far as Irkutsk. From the railroad train one can see the long line of the old topographic surface rising higher and higher in the mountains toward the south, and an observer familiar with the features and interpretation of the Appalachian mountains can not doubt that he has here in northern Asia a warped peneplain, which, like the Schooley peneplain, is somewhat extensively dissected.

The preceding observations may all of them be said to be marginal to the great highlands of central Asia, and although the Peï-t'ai surface lies at an altitude of 10,000 feet, 3,000 meters, it may be considered a daring proposition to extend an inference regarding the peneplain to the highest

* La Face de la Terre, vol. III, p. 145

plateaus of the world.   But we need not rest upon mere inference.   Davis
has described the summit character of the various ranges of the Tién-shan.*
He says:

> Certain observations made in the central and northern ranges [of the Tién-shan] near
> lakes Son-kul and Issak-kul, and on the steppes that border the mountains on the north,
> led to the belief that the region had been very generally worn down to moderate or small
> relief since the time of greater deformation, which probably occurred in the Mesozoic age;
> that large areas of subdued or extinguished mountain structures are still to be seen in
> the low ranges and in the steppes north of the Ili river, and that the present relief of
> many of the higher Tién-shan ranges is the result of a somewhat disorderly uplift and of a
> more or less complete dissection of dislocated parts of the worn-down region.   Mr. Hunt-
> ington's report shows the application of these conclusions to a large part of the central and
> southern Tién-shan.

We have no direct evidence of the age of this peneplain, which is now
elevated to altitudes approaching 14,000 feet, 4,200 meters, but the perfec-
tion of the profile of the Bural-bas-tau and other ranges sketched by Davis
suggest that they have not long been exposed in their present altitude.
And when we consult the geologic section of the adjacent region of western
Turkestan, we find no record earlier than the middle Tertiary of voluminous deposition, such as the mountains are capable of yielding.

The descriptions of Mongolia and northern Tibet give the physiog-
rapher no reason to expect that an ancient topographic surface may there
survive.   Much of the region is buried beneath Pliocene and Quaternary
sediments, and the deeply sunken surface of the hard rocks is hidden from
view.   The mountain ranges which rise above the sands have been sharply
sculptured and appear to be the skeletonized edges of warped and tilted
blocks.   While it is possible that Davis's observations of the Tién-shan
may be repeated elsewhere in the Kuen-lung or the Nan-shan, it is more
probable that our own experience in the southeastern extension of these
ranges, the Ts'in-ling-shan, will be paralleled, and that the observer will be
able to recognize nothing older than a mature surface of late Tertiary date.

When we consider that physiographic studies have been applied to the
interpretation of the features of Asia only within the last three years, it
arouses some surprise to find so much evidence of a surviving peneplain;
but that evidence appears to deserve frank recognition in view of the series
of events leading up to the early Tertiary, the stratigraphic record of the
Cretaceous and Tertiary, and the character of plateaus and plains of denudation which so much of the land presents, even at high altitudes.

It is, however, improbable that an entire continent of such extent as
Asia should be completely peneplained.   In America the peneplanation
of the Cretaceous period failed to reduce the Unaka mountains of North

---

* Explorations in Turkestan, Carnegie Institution of Washington Publication No. 26, p. 72.

Carolina* and the case being considered generic it was proposed that an extended group of heights surviving above a peneplain should be called a "unaka." It is highly probable that one or more unakas will be found in Asia.

The dissection of the Peï-t'ai peneplain (using that name tentatively as a general term) has proceeded differently in different regions. Where the plain is buried beneath Tertiary deposits, as on the Amur, it may be said to be intact. Where it has been slightly warped or elevated to a moderate height over vast areas, as in northern Siberia, it is still clearly recognizable in the featureless plateaus. It may also be seen as a general line of some mountain profiles where its surface is more steeply tilted, as in the northern Altai. It is found as a mere remnant in such summits as the Bural-bas-tau and Peï-t'ai. It is no longer to be seen in districts such as Shan-tung, where early Tertiary faulting gave rise to acutely accented relief; nor in the Ts'in-ling-shan, where warping in mid-Tertiary time occasioned the development of a mature but hilly landscape.

Throughout North China we can distinguish three phases of physiographic development later than the Peï-t'ai plain: the T'ang-hién, one of mature erosion; the Hin-chóu, one of aggradation; and the Fön-ho, one of mountain growth. In central China we recognized but two: the Ts'in-ling, which corresponds to the first two; and the Yang-tzï, which is probably closely equivalent to the Fön-ho.

The distinction between the T'ang-hién and the Hin-chóu epochs depends upon the accumulation of the early deposits of loess, which covered a vast area, but an area conditioned by depression and geographic position with reference to streams flowing from the basins of central Asia. These conditions were by no means universal; they were bounded on the south by the hill district of the Ts'in-ling-shan; and it is probable that outside the region of their extent the distinction between the epochs can not be made. The equivalent phase of mature topography, which we have called the Ts'in-ling phase, will then be found where any feature older than the Fön-ho or Yang-tzï stage is recognizable.

The Ts'in-ling stage of topographic development characterizes the summit views of the Ts'in-ling-shan, the mountains of the Han valley, and the ranges of the middle Yang-tzï region, where we traversed them from the Weï valley to I-chang on the Yang-tzï. It is marked by more or less decided mature relief, above well-developed valley floors, in which the younger canyons are cut. It presumably succeeds the Peï-t'ai peneplain, though the earlier existence of the latter in this region can not be demonstrated, since erosion progressed too far in the Ts'in-ling cycle. *It certainly replaces the tectonic relief which resulted from the Permo-Mesozoic folding.*

---

* Geomorphology of the Appalachians, Hayes and Campbell, Nat. Geog. Soc. Mag., vol. VI, p. 63, 1894.

The last statement runs contrary to the views of many geologists, and particularly to the views of those who hold with the master, Suess, that the present mountain ranges are a direct consequence of the compressing forces that folded the strata within them. There are, indeed, two distinct points of view, each of which is the outcome of observation in a region that by its special character determines the inference. With a few exceptions, European geologists are dominated by the stupendous structures of the Alps, and have neglected physiographic studies. American observers twenty years ago discovered a new line of interpretation in physiography, and applying it first to the Appalachians have since extended it to other ranges and continents. In the light of present knowledge it appears safe to generalize as follows: *Those mountain chains which exhibit folded structures of post-Eocene development owe their elevation in part possibly to the original effects of that compression*, and in part to subsequent efforts of a force acting in the same sense, but producing upwarps and downwarps that are independent of anticlines and synclines, yet related in general position and trend to the folded chain as a whole. On the other hand, *altitudes due to folded structures of the Permo-Mesozoic or older epochs of diastrophism were long since planed away by erosion*, and though the structures may be involved in relatively modern upwarps, they are not related to the existing elevation. This is true in spite of the general fact, inherent in the broader continental and oceanic features of the earth, that some zones of orogenic activity retain their dynamic character from an early geologic date to the present, as witness the Wu-t'ai-shan.

There are thus two types. Of the former, characterized by post-Eocene folding and later warping in the same sense, the Karpathians are the best known example,* the structure and physiography of the Alps being too obscure to serve as a type. The second, characterized by Fermo-Mesozoic folding, peneplanation, and relatively recent warping, is represented by the classic Appalachians.

In a region where the relief is directly due to anticlinal elevations and synclinal depressions the relation between the altitudes and the structures must be such that the anticlinoria at least correspond to heights and the synclinoria to lows. A case in point is that of the Lewis range in Montana,† where the altitudes bear these relations. But these relations do not hold for the mountains of Central China; their heights and lows are related to upwarps and downwarps, which are not coincident with the complex and intruded structures of Permo-Mesozoic time, but are everywhere sculptured

---

*European Studies, B. Willis, Carnegie Institution of Washington Year Book No. 4. Bau und Bild der Karpaten, by V. Uhlig, in Bau und Bild Oesterreichs, Wien, 1903.

†Stratigraphy and Structure of the Lewis and Livingston Ranges, B. Willis, Bull. G. S. A., vol. XIII, p. 346.

with the mature erosion forms of the Ts'in-ling stage. And strictly according to the later warping and faulting are the depths of the young canyons, which pursue courses that are in some measure consequent on the slopes of recent upwarps, but in general inconsequent to the older structure. The present relations of altitudes and folds show that tectonic elevations of Permo-Mesozoic age have given place to warped elevations of modern age; and the physiographic record makes it clear that one or more periods of little or no elevation intervened.

The present topographic cycle, which we may call either the Fön-ho or the Yang-tzï cycle, is characterized by the development of basins, warped surfaces, fault-scarps, and great mountain ranges, and consequently also by youthful erosion forms and extensive plains of aggradation. The scarp, the canyon, and the alluvial fan are the marked accents of the cycle and are the noted features of Asiatic topography. The warped surface that bears the sculpture of a preceding cycle is also of general occurrence, but has less often been noted.

The map, plate 8, comprising the Chinese Empire and parts of adjacent regions, represents the distribution of elevated and depressed areas. It is based on actual altitudes and is a rough hypsometric map; but it is not accurate in that sense, because, on the one hand, the number of points whose altitude is known is inadequate to accuracy, and on the other, effects of erosion have been disregarded intentionally. The surface which the contours represent constitutes the basins, slopes, ranges, and plateaus as they would exist if a plain had been warped and dislocated and had not been eroded. The map thus expresses the hypothesis that Asia was reduced to a general peneplain and has since been warped in a manner to produce differences of elevation exceeding 6,000 meters, 20,000 feet. In the preceding pages and in volume I, part I, of this publication I have given the facts which indicate that, in the districts we observed, the warping has taken place chiefly during late Tertiary and Quaternary times. My present purpose is to extend the inferences regarding modern mountain growths to regions that are genetically related to those which we saw.

First, with reference to the region of the great alluvial plain of eastern China, it is usually recognized that it is an area of depression, a downwarp extending from northern Manchuria to Hu-nan and filled with alluvium of the great rivers that debouch from the western and southern mountains. We have shown that Shan-tung, the peninsula of the Eastern Mountains, has since the episode of the early Tertiary faulting stood as a horst in the sinking region, its margins being bent down but its interior not notably raised in the process.*

---

* Research in China, vol. I, part I, page 83.

The eastern margin of this downwarp lies in the Yellow Sea, possibly between the point of Korea and Shanghai. The mountains of southeastern China may be said to limit it on the southeast. The Yang-tzï defines it northeast from the Tung-ting lake. The hills of An-hui and Hu-peï appear to be within its area.

The western margin of the downwarp is the slope of the mountains, comprising the Khingan range in the north, the so-called plateau of Shan-si in a mid-stretch, and the mountains of eastern Hu-peï further south. Von Richthofen described the limit, between the plain and the mountains as a fault, which appears on his map, Versuch einer Tektonischen Karte des nördlichen Chinas, as the Khingan Linie.* He does not cite any evidence of faulting on the line itself, it being drawn indeed in the plain of alluvium; and according to our observations on three different sections the passage from the plain to the mountains is a zone of warping, not a line of dislocation. Where, in latitude 49°, the Siberian railroad descends to the Sungari, the eastern slope of the Khingan is a tilted, dissected, but unbroken peneplain. Where the Sha-ho, in latitude 39°, has cut its autogenous valley, and the hills about T'ang-hién lie half-buried in the plain, the effects of modern warping are obvious. Normal faulting, though present in the Ning-shan basin, occurred at a remote Tertiary date, and erosion has reversed the relief to which it gave rise. Again, in latitude 31°, where the Yang-tzï emerges from its profound gorges at I-chang, the mountain slope that faces the far-spreading river plain is a tilted surface of erosion, showing a continuous stratum of Carboniferous limestone, which toward the base is overlain by the K'ui-chóu red beds in appropriate stratigraphic sequence. It is a warped surface, not a fault; and there is no evidence that it is limited by a fault at the base.

This zone of warping was crossed by von Richthofen along two routes of travel in Shan-si; the one in latitude 33°, northwest of "Hwai-king-fu;" the other in latitude 38°, on the great road from T'ai-yüan-fu to the eastern plain. In the former traverse he ascended to the plateau over a monoclinal flexure on an erosion surface that is sculptured on Carboniferous shales and Sinian limestones† and that forms spurs between deep gulches. The steep slopes of the spurs end in a line and thus simulate a fault-scarp, but the structure is that of a simple flexure and is so described by the traveler. At the more northerly crossing von Richthofen observed step-faulting which he describes as follows:‡

After having observed the structure of the mountains in southern Shan-si, it was for me a matter of no slight interest to see in what manner the horizontal strata of the plateau might break away or sink toward the Great Plain. As we have seen, the plateau

---

*China, vol. II.   † Ibid., vol. II, p. 407.   ‡ Ibid., vol. II, pp. 440–442.

ended on the south with a short monoclinal flexure. Were that to be the case in this region the plateau must extend far toward the east; but this did not seem probable, as according to all accounts I was near the end of the anthracite region. In fact I came sooner than I expected to the first fault. The general relation was indeed difficult to understand in a single rapid traverse, but it was soon clear that the plateau fell off in steps, which are occasioned by faults along a great fault zone, which runs from north to south and is in part connected with gentle flexures. If one regards the attitude of the strata in individual cases, one often finds it confusing, for locally the limestone dips in various directions and with different angles. This is particularly noticeable in the descent from each separate step. But if one looks back toward the west from a little distance and regards the steep slope of the ridge which has just been crossed, one is surprised to observe in general only horizontal lines of stratification. In every such case one is on that portion of the surface which has sunk down with reference to the higher steps. The variations in the altitude of the strata may possibly be related in part to the displacement of the beds in the tremendous process of faulting, or may in part be due to caving in, there being in some places caverns, which occur especially in the Rauchwacke.

While these observations are avowedly incomplete and do not carry conviction to the reader, the conclusion is not inconsistent with the warping observed elsewhere. Warping and faulting are related phenomena; the latter is an extreme effect of a concentrated stress set up in consequence of the former, and may be looked for wherever the zone between an area of depression and one of elevation is relatively narrow as compared with the difference of altitude resulting from the movement.

I conclude that the weight of evidence shows that the passage from the depressed region to the elevated plateaus of the mountainous area is by a warped surface more generally than by a dislocation, and that the "Khingan Linie" is to be regarded as a zone of monoclinal flexure, not as a fault.

The depressed region just described is one of the eastern or outer provinces of the continent. While it has subsided somewhat, the region on the northwest has risen considerably more. The physiographic record shows that up to the close of the Hin-chóu epoch the difference of altitude between the two was no more than that of a gentle, continuous slope, but since that time it has been increased to 6,000 or 7,000 feet, 1,800 or 2,100 meters, on a moderate allowance of 5,000 feet for the elevation of the T'ang-hién surface in the plateaus of Shan-si, and of 1,000 to 2,000 feet, 300 to 600 meters, for the depression of the same surface beneath the Great Plain.

This difference in altitude has developed within the late Pliocene and the Quaternary; probably chiefly during the latter. There are two conceptions of the manner: The region which is high may be regarded as a horst, which has stood firm while the adjacent area has sunk down below its level; and simultaneously the ocean basins have so deepened that sea-level has been lowered 5,000 feet, 1,500 meters. Or, there have been both positive and negative movements—not sinking only, but rising also—and consequently readjustments of sea-level and of altitudes of various areas.

The element of time is important in the hypothesis which ascribes all epeirogenic changes of level to subsidence in obedience to gravity; but the movements in Asia are so recent that time fails. The physiographic history of the mountains of Shan-si is plain and consistent. The young slopes and scarps and canyons are obvious facts. However we may differ in a reasonable estimate of their possible age, we can not assign to them an antiquity such that during their development sea-level might have sunk 5,000 feet, 1,500 meters. To do so would not only contradict the direct evidence of comparison with topographic features elsewhere, which are definitely dated by relation to late Tertiary sediments, but would require explanation of the absence of a marine record where there is an erosion record instead. This question enters into the study of the highlands of Asia as a fundamental distinction between two very different interpretations of the orogeny and epeirogeny of the continent. My views are more fully stated in the following chapter on the hypotheses of continental structure. It suffices here to state that I hold it to be true that when some masses sink notably, whether in continental or oceanic regions, other masses rise notably; while some parts of Asia have certainly subsided during the latest movements, other parts have risen, and the upward movements, which are measured by many thousand feet, are greatly in excess of the downward movements within the continent.

The mountainous region of northwestern China may be said to consist of the Khingan range north of latitude 40°, where that height is the margin of the Mongolian plateau simply, and of the various ranges of Chï-li and Shan-si, which as a group are distinct from the Mongolian plateau. Among the latter are the Nan-k'ou range northwest of Peking, the Wu-t'ai-shan, and the plateau of Shan-si with the Ho-shan. The physiographic history of these mountains and the reasons for believing them to have been elevated during the Fön-ho epoch of the Quaternary have been sufficiently set forth.

Northwest of the mountain system of Chï-li and Shan-si is a depressed zone characterized by the basins of Ta-tung-fu, latitude 40°; of Hin-chóu, latitude 38° 30'; of T'ai-yüan-fu, latitude 37° 30'; of P'ing-yang-fu, latitude 36°; and of the Weï valley, latitude 34° 30'. The last stretches far to the west, north of the Ts'in-ling-shan, and is related to the depression of Kan-chóu and Sü-chóu, which lies along the northeastern border of the Nan-shan. Von Richthofen describes the basin of Ta-tung-fu.* We refer to

---

*China, vol. II, p. 359.

our own account of the others in volume I, except with reference to the most northwestern, for which Lóczy may be consulted.*

The zone of these basins, comprising an arc which surrounds Mongolia and Ordos on the southeast and which is more than 1,500 miles, 2,400 kilometers long, is a downwarp with reference to the mountain regions that adjoin it. It is related to the heights southeast and south of it by warped surfaces, which are, however, faulted throughout considerable stretches, producing great scarps several thousand feet high that face inward toward central Asia. The ascent toward the northwest or north is generally by long, gently inclined slopes, but locally by more steeply tilted surfaces, and occasionally by a fault. The O-shan fault in central Shan-si, which defines the Fön-ho graben on the west, is the only one which has been definitely recognized on that side.

In its internal displacements the zone of depression is not simple. The basins lie *en échelon* and are separated by moderate upwarps that traverse the zone diagonally or directly as the case may be. The dividing ridges may be enumerated as follows: the Man-t'o-shan, separating the basins of Ta-tung-fu and Hin-chóu; the Shï-ling, south of the latter; the Si-yau-ling and Si-sin-ling in central Shan-si, between the T'ai-yüan-fu and P'ing-yang-fu basins; and the lesser upwarp dividing the Fön-ho from the salt lake basin in southern Shan-si. These details of warping are characterized by the youthful features of the Fön-ho epoch, namely, the mantle of Huang-t'u formation not yet removed, the relation to antecedent streams, and the characteristic deep gorges.

This zone is one of the major structure lines of the continent, which is recognized in modern displacements by relatively slight elevation above sea. It agrees in general trend with the directions of folding of pre-Cambrian as well as of Fermo-Mesozoic deformation. It was apparently the littoral of pre-Cambrian seas, a strip bordering the downwarps in which the Wu-t'ai and Hu-t'o sediments accumulated to great thickness; but after they were folded it became part of the land across which the Sinian sea transgressed and over which Carboniferous continental deposits more or less generally accumulated. During the long quiescence of the continent it does not appear to have been a distinctive line; in the Fermo-Mesozoic disturbance it was folded; and in the recent extraordinary diastrophism it has again become manifest in marked differences of elevation.

Mongolia lies northwest of the basins of Shan-si and north of the depression of Shen-si and Kan-su. It is a region throughout which separate ranges rise from great expanses of Pliocene and Quaternary desert waste, a region in which the evidences of structure and of physiographic development

---

* Reise des Grafen Széchenyi, vol. I, p. 499 *et seq.*

are disconnected. They have not been observed from the point of view we are now taking. The available data relating to the structural trends, the isolated ridges, the desert plains, and the deep depressions are gathered by Suess into a masterly description in the third volume of Das Antlitz der Erde. The acute features of dislocation and erosion; the warping of the Tertiary Gobi deposits and their occurrence high on the ranges from which they extend to basins below sea-level: these suffice to bring the period of diastrophism to which the relief is due within the later Tertiary and Quaternary. I quote Huntington, who has studied the western part of the vast region.* In regard to the Zorabad basin, he says:

> Apparently it was first occupied by the sea and later became dry land. Then, by the warping of the earth's crust, it was converted into a lake, which in time was drained by the cutting of a gorge. As the water of the lake receded, gravel was washed in from the sides and covered the lake deposits. Since that time the gorge at the outlet has been cut deeper, the various deposits have all been more or less dissected, and terraces have been formed. At intervals during the progress of these events, warping has gone on in such a fashion that the size of the basin has continually diminished and all the deposits except the most recent gravels have been warped along the edges, although apparently remaining horizontal in the center of the basin. Most of this history probably belongs to Tertiary times, although the dissection of the lake deposits and the formation of the terraces almost certainly belong to the present geological era.
>
> In order to understand the geological history of Persia it will be necessary to ascertain to what extent a similar series of events has occurred in other basins. What few facts are known indicate that the history of all the basins is similar to that of Zorabad, with the exception of the lake episode. The only lakes of which we have record in the other basins occurred at a later time and were due to changes of climate rather than to warping of the crust.

The mountains of Trans-Baikalia constitute a group which is clearly distinguished from the Gobi region on the south, the plateaus to the northeast, and the plains of the amphitheater of Irkutsk to the northwest. Whereas the Gobi is a region of displacement and aggradation and the plateaus are one of regional uplift and the plains one of regional depression, Trans-Baikalia is characterized by elevation, faulting, and denudation. I have already cited the evidence adduced by Suess for ascribing to the basin of Lake Baikal a late Tertiary date, and stated that the conclusion may be extended to the ranges of Trans-Baikalia. This applies to the principal heights in the northeastern district. It is doubtful how far the Tertiary uplift may be traced in the adjoining plateau district, where the sharply defined canyon of the Lena has the aspect of a Quaternary gorge. I am inclined to regard the Trans-Baikal mountains as an insular faulted upwarp, like Shan-tung, and to infer that the plateau region is of later elevation.

*Explorations in Turkestan, Carnegie Institution of Washington Publication No. 26, p. 242.

From Siberia we may return to central China.

The Ts'in-ling-shan rises from the valley of the Weï by a steep warped surface and fault-scarp, which we ascribe to the Quaternary (Fön-ho, or Yang-tzï) epoch. Among the heights of the range we recognize a mature surface, which we attribute to' the Ts'in-ling cycle of late Tertiary time. It extends beyond the Ts'in-ling range, across the Han watershed to the middle Yang-tzï region, and is strongly warped. The great features to which the warping gives rise are four: the Ts'in-ling-shan, the Han down-warp, the Kiu-lung-shan between the Han and the Yang-tzï, and the present basin of Ssï-ch'uan.

The Ts'in-ling-shan is a long upwarp, the eastern continuation of the Kuen-lung; it attains a general elevation of 7,000 to 9,000 feet, 2,100 to 2,700 meters, and in the special accent, the Ta-pai-shan, rises to 12,000 feet, 3,600 meters. On the south it sinks to the Han downwarp, which is a broad depression having summit altitudes of about 3,000 feet, 900 meters, which are modified by local upwarps that probably reach 6,000 feet, 1,800 meters, and by two pronounced downwarps, those of Han-chung-fu and Hing-an-fu. The floors of these basins are plains of aggradation, that of Han-chung-fu approximately 1,500 feet, 450 meters, that of Hing-an-fu very nearly 800 feet, 240 meters, above sea; but the Ts'in-ling erosion surface sinks beneath the Quaternary deposit, being depressed to a still lower level. At Hing-an-fu the coarse gravels, sands, and clays form bluffs along the Han. The Red Basin of Ssï-ch'uan is a downwarp like the two mentioned, except that it is very much larger, and, as the remarkable delta plain of Chung-t'ing-fu shows, is a region of continued subsidence and aggradation. The Kiu-lung-shan is an upwarp containing summit altitudes of 8,000 to possibly 12,000 feet, 2,400 to 3,600 meters. It lies between the basin of Ssï-ch'uan which is on the west and the great downwarp of the lower Yang-tzï, its warped surface passing beneath the flood-plain at I-chang. On the north it is bounded by the Han downwarp. Southward it extends across the Yang-tzï, its surface declining to depressions that are not yet observed.

The warped surface of Central China, which may thus be analyzed, exhibits differences of altitude of 12,000 feet. It is, so far as we saw it and we believe generally, an erosion surface which, while retaining notable relief, assumed mature features by the close of the Tertiary period. It may have persisted at a low elevation into Quaternary time. During the Quaternary it has been warped, giving rise to the great mountain chains and extensive basins that now exist. The warping characterizes the Yang-tzï epoch of erosion, and the youthful aspect of the canyons of that cycle is the ground for the inference that the movement did not assume notable proportions until Quaternary time.

The elevation of the Ts'in-ling-shan is continued toward the west by the Kuen-lung and the Nan-shan, which rise to much greater heights; but the physiographic aspect appears to change. The broad upwarp which, in longitude 108° east, is 100 miles, 160 kilometers, across without a break, is replaced by several ranges separated by valleys and basins. The effects of warping are more complex and are probably accented by faulting. The type is rather that of the Gobi than that of central China.

Tibet and the Himalayas finally claim attention.

The Alps of eastern Tibet, the Yung-ling, tower above the lowlands of Ssï-ch'uan in wonderful grandeur to an extreme altitude of 7,000 meters. We have no definite knowledge of the physiographic aspects of this great mountain face, but in one respect it differs markedly from the southern front of the Himalayas, with which it may naturally be compared. The rivers that drain the Himalayas flow directly across the range, after the manner of consequent streams, which have developed at right angles to the trend and been extended by headwater erosion in consequence of a combination of favoring conditions. The rivers of the Tibetan Alps, on the contrary, flow southwest between high ranges, which direct them in courses diagonal to the lines of shortest descent toward the basin of Ssï-ch'uan. They thus have the character of streams which are consequent upon a folded or faulted surface and take their own way down the axial lines of the major depressions. There is thus reason to regard this slope as being composed of successive upwarps or fault-blocks, which lie *en échelon* and sink at their southeastern ends to the lowland of Ssï-ch'uan. The character is expressed in the contours which define the slope north of latitude 30° and about the meridian of 103° east.

Toward the west and south the great ranges of the Tibetan Alps are bounded apparently by the broad plateau, which is deeply incised by the canyons of the upper Yang-tzï and the Mekong and their several branches. The parallelism of these great rivers west of the one-hundredth meridian may be an effect of unknown tectonic lines, but it is with equal reason explicable as the growth of autogenous canyons on a uniform slope. The meanders of the Yang-tzï in latitude 25° north may be attributed to capture across a fault-scarp.

The western part of the map includes the vast highland of Tibet and its bounding ranges, the Altin-tagh on the north and the Himalayas on the south. The structural character of the plateau is not known, but it appears to be that of a mass which has been forced above a position of equilibrium and which has consequently broken into blocks that have suffered diverse displacements. The depressions are deeply filled with Pliocene and Quaternary sediments.

The limiting ranges hold apparently homologous relations to the plateaus; but they differ greatly in history and structure. The Altin-tagh, occupying part of the northern Tethys, was folded primarily in the Fermo-Mesozoic revolution, and its essential structures thus date from a somewhat remote time. The Himalayas did not suffer compression till the beginning of the period of diastrophism that marks the late Tertiary and Quaternary. The Altin-tagh seems to rise from the Tarim basin as a continuation of the eroded surface which plunges beneath the Gobi deposits of the desert; it resembles apparently the warped surface by which the Shan-si mountain region sinks beneath the plains of eastern China; and though possibly in some sections broken by normal faults, it does not differ from other mountain slopes of Asia. The Himalaya range, on the other hand, is unlike them. It is separated from the plains at its southern base by thrusts of great magnitude, which dip beneath the upraised mass. The effect is as if the range were pushed southward—were overthrust; it may equally well be expressed by the statement that the lowlands are pushed northward—are underthrust. Since the overthrusting or underthrusting, whichever is the dominant fact, involves movement on an inclined plane, the mechanical condition is that which would result from driving a wedge under the range. Either the range must be raised or the wedge must be depressed, or both movements may occur. In discussing the analogous case of the Lewis range of the Rocky Mountains in Montana, I have shown that an elevation of 3,400 feet is probably attributable to a displacement of 7 miles on the flat thrust which underlies the Algonkian strata.* The thrusts beneath the Himalayas are apparently even more important factors in the relative elevation of the mountain mass. Among Asiatic mountains the Himalayas thus present a unique case of mechanical relations.

The elevation of the Tibetan plateau (Isle Tibet) is apparently an effect of the underthrusting, to which we may attribute some considerable part of the altitude of the Himalayas. I conceive that the plateau is the surface of a deep-seated, strongly compressed sub-Tibetan mass.

---

*Stratigraphy and Structure, Lewis and Livingston Ranges, Montana, Bull. G. S. A., vol. XIII, p. 345.

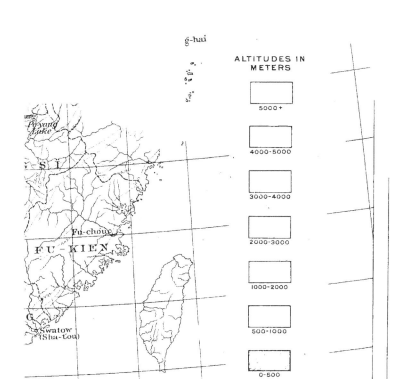

g-hai

ALTITUDES IN
METERS

5000+

4000-5000

3000-4000

2000-3000

1000-2000

500-1000

0-500

118° 122° 124° 126°

Pi-yang
Lake

-SI

Fu-chou

FU-KIEN

G.
Swatow
(Sha-tou)

m Stielers Atlas, IS
and 64 by C.Bari
s compiled from al
d sources by A.O.L

Structure and physiography
interpreted by Bailey Willis

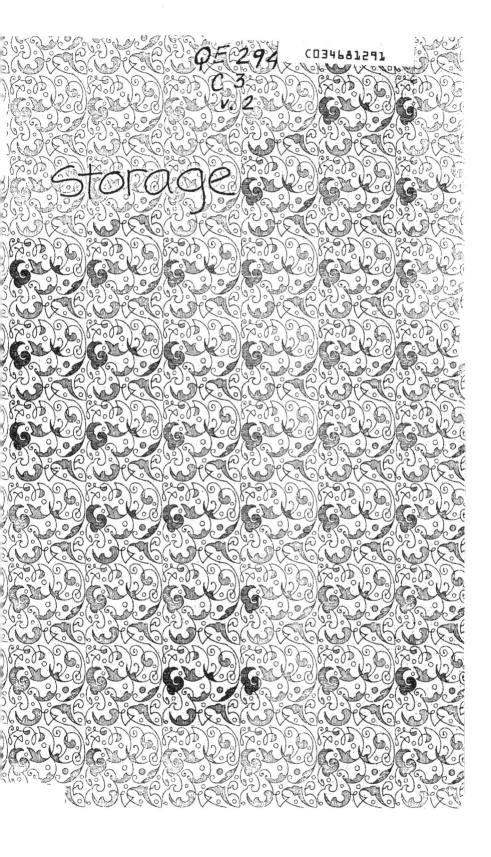